Urban Spaces
No. 5

Featuring Green Design Strategies

城市空间与景观设计 5

[美] 约翰·莫里斯·狄克逊　编著

中国建筑工业出版社

著作权合同登记图字：01-2008-3795号

图书在版编目（CIP）数据

城市空间与景观设计5/（美）狄克逊编著；—北京：中国建筑工业出版社，2008
　ISBN 978-7-112-10277-8

　Ⅰ.城... Ⅱ.狄... Ⅲ.①城市空间-建筑设计②城市-景观-设计　Ⅳ.TU984.11

中国版本图书馆CIP数据核字（2008）第124229号

Copyright © 2007 Visual Reference Publication, Inc.
All rights reserved.

本书由美国VRP出版社授权出版

责任编辑：程素荣

城市空间与景观设计 5
[美] 约翰·莫里斯·狄克逊　编著
*
中国建筑工业出版社出版、发行（北京西郊百万庄）
各地新华书店、建筑书店经销
北京嘉泰利德公司制版
北京盛通印刷股份有限公司印刷
*
开本：965×1270毫米　1/16　印张：19　字数：700千字
2009年1月第一版　2009年1月第一次印刷
定价：260.00元
ISBN 978-7-112-10277-8
　　　（17080）

版权所有　翻印必究
如有印装质量问题，可寄本社退换
（邮政编码 100037）

Contents

Contents by Project Type	4
Preface by Richard M. Rosan, FAIA	7
President: Urban Land Institute Worldwide	
Introduction by John Morris Dixon, FAIA	9
Ankrom Moisan Architects	17
annex 5	25
Architects Orange	33
BALDAUF CATTON VON ECKARTSBERG ARCHITECTS	41
Beeler Guest Owens Architects	49
Booth Hansen	57
Braun + Yoshida Architects	65
Carter & Burgess, Inc.	73
CMSS Architects, PC	81
Derck & Edson Associates	89
Duany Plater-Zyberk & Company	97
Ehrenkrantz Eckstut & Kuhn Architects	105
Hughes, Good, O'Leary & Ryan, Inc.	113
James, Harwick+Partners, Inc.	121
Joseph Wong Design Associates (JWDA)	129
JPRA Architects	137
LandDesign	145
Lessard Group Inc.	153
Lucien Lagrange Architects	161
MBH	169
Moule & Polyzoides	177
MSI	185
MulvanneyG2 Architecture	193
NewmanGarrisonGilmour+Partners	201
O'Brien & Associates	209
Perkowitz+Ruth Architects	217
RLC Architects, P.A.	225
RTKL Associates	233
Sasaki Associates, Inc.	241
SmithGroup	249
Starck Architecture + Planning	257
Studio 39 Landscape Architecture, PC	265
Thompson, Ventulett, Stainback & Associates (TVS)	273
Torti Gallas and Partners	281
Project Credits	289
Index by Project	302

Contents by Project Type

Communities/Neighborhoods
(typically with wide mix of uses, on cleared land, with new buildings and infrastructure; see also Urban Redevelopment)
Alys Beach, Walton County, FL, **99**
The Architect Collection at Stapleton, Denver, CO, **70**
Artessa at Quarry Village, San Antonio, TX, **54**
Arverne-by-the-Sea, Arverne, NY, **108**
Battery Park City, New York, NY, **106**
Beacon Point at Liberty Station, San Diego, CA, **258**
Belle Creek, Henderson, CO, **66**
Belmar Row Houses, Lakewood, CO, **68**
Bridge Street Town Centre, McKinney, TX, **224**
The Brownstone Collection at Highlands Ranch Town Center, Highlands Ranch, CO, **69**
Cool Springs Mixed-Use Development, Franklin, TN, **152**
East Fraserlands, British Columbia, Canada, **102**
Ibn Battuta Mall Phase II, Dubai, UAE, **78**
Kincora, Loudoun County, VA, **84**
Lansdowne Village Greens, Loudoun County, VA, **277**
Magnolia, Charleston, SC, **114**
Marineland, Flagler County, FL, **104**
Piemonte at Ontario Center, Ontario, CA, **36**
Schooner Bay, Great Abaco, Bahamas, **100**
Seaside, Walton County, FL, **98**
Sky, Calhoun County, FL, **101**
Sonoma Mountain Village, Rehnert Park, CA, **170**
The Triangle, Austin, TX, **56**
Upper Rock District, Rockville, MD, **103**

Convention Center
Washington Convention Center, Washington, DC, **282**

Cultural Facilities
The China Film Institute, Beijing, China, **236**
China Science and Technology Museum, Beijing, China, **28**
Georgia Aquarium, Atlanta, GA, **288**
The Joffrey Tower, Chicago, IL **64**
New Jiang Wan Cultural Center, Shanghai, China, **234**
Plaza Norte, Santiago, Chile, **286**
Reginald F. Lewis Museum of Maryland African American History and Culture, Baltimore, Maryland, **239**
Workhouse Art Center at Lorton, Lorton, VA, **266**

Educational Facilities
Bridge Street Town Centre, McKinney, TX, **22**
North County Regional Education Center, San Marcos, CA, **131**
Technology Square, Georgia Institute of Technology, Atlanta, GA, **284**

Entertainment Complexes
Bella Terra, Huntington Beach, CA, **221**
Galerias Hipodromo, Tijuana, Mexico, **219**
Hollywood & Highland, Los Angeles, CA, **112**
Las Colinas Entertainment, Irving, TX, **76**
Palladium, Birmingham, MI, **142**
Plaza Norte, Santiago, Chile, **286**
Shenzhen Bay Seafront Urban Design, Shenzhen, China, **150**
The Shops at Highland Village, Highland Village, TX, **214**

Government Complexes
Cook County Circuit Courthouse, Chicago, IL, **58**
Redmond City Hall, Redmond, WA, **194**

Hotel/Resort/Conference Centers
Allen Plaza, Atlanta, GA, **116**
Bridge Street Town Centre, McKinney, TX, **224**
Celebration Hotel, Celebration, FL, **188**
Fudan Crowne Plaza Hotel, Shanghai, China, **200**
Hard Rock Hotel, Chicago, IL, **166**
Palmolive Building, Chicago, IL, **62**
Sofitel JJ Oriental Hotel, Shanghai, China, **133**
The Towne of Seahaven, Panama City Beach, FL, **148**

Mixed-Use Developments
(includes substantial mix of uses, beyond accessory uses, such as parking or retail in otherwise office or residential buildings)
3030 Clarendon Boulevard, Arlington, VA, **274**
Alexan Pacific Grove, Orange, CA, **40**
Arterra, Reno, NV, **24**
Beijing CBD Office Park, Beijing, China, **32**
Beijing Century City East, Beijing, China, **30**
Boca Raton Mixed-use, Boca Raton, FL, **80**
Bridgeport Condominiums, Portland, OR, **20**
Broadway Grand, Oakland, CA, **173**
Centergate at Baldwin Park, Orlando, FL, **278**
City Crossing, Shenzhen, China, **240**
City Lights, Dallas, TX, **210**
Cityville Greenville, Dallas, TX, **122**
Deep Blue Plaza, Hangzhou, China, **134**
Elleven, Los Angeles, **24**
The Ellington, Washington, DC, **274**
Gregory Lofts, Portland, OR, **22**
Highland Park, Washington, DC, **275**
Hollywood & Highland, Los Angeles, CA, **112**
Ibn Battuta Mall Phase II, Dubai, UAE, **78**
The Joffrey Tower, Chicago, IL **64**
Kenyon Square, Washington, DC, **275**
Lane Field, San Diego, CA, **132**
Luma, Los Angeles, **24**
Martin Luther King Plaza, Philadelphia, PA, **278**
Melrose Triangle, West Hollywood, CA, **223**
Mercato, Bend, OR, **47**
Midtown Reston Town Center, Reston, VA, **154**
Military Family Housing, Fort Belvoir, VA, **278**
Museum Place, Fort Worth, TX, **127**

New River at Las Olas, Fort Lauderdale, FL, **109**
Odeon Union Square, San Francisco, CA, **174**
One South Market, San Jose, CA, **24**
Park Place Condominiums, Washington, DC, **275**
Park Triangle, Washington, DC, **275**
Piemonte at Ontario Center, Ontario, CA, **36**
Ponto Beachfront Village, Carlsbad, CA, **262**
Rollin Street, Seattle, WA, **23**
Salishan, Tacoma, WA, **278**
Shirlington Village, Arlington, VA, **276**
The Towne of Seahaven, Panama City Beach, FL, **148**
Twinbrook Commons, Rockville, MD, **276**
Zhejiang Fortune Financial Center, Hangzhou, China, **196**

Office Buildings/Developments
951 Yamato, Boca Raton, FL, **230**
Allen Plaza, Atlanta, GA, **116**
Beijing CBD Office Park, Beijing, China, **32**
Beijing Century City East, Beijing, China, **30**
City Crossing, Shenzhen, China, **240**
Columbus Riverfront Office Building, Columbus, GA, **118**
Deep Blue Plaza, Hangzhou, China, **134**
Discovery Communications Headquarters, Silver Spring, MD, **252**
Domus Office Tower, Hallandale Beach, FL, **232**
Highmark Data Center, Harrisburg, PA, **238**
Kincora, Loudoun County, VA, **84**
Lititz Watch Technicum, Lititz, PA, **90**
MB Financial Bank, Chicago, IL, **60**
The Robert Redford Building for the Natural Resources Defense Council (NRDC), Santa Monica, CA, **178**
U.S. Epperson/Lynn Insurance Group Corporate Headquarters, Boca Raton, FL, **228**
Visteon Village Corporate Headquarters, Van Buren Township, MI, **254**
Zhangjiang Semiconductor Research Park, Phase II, Shanghai, China, **198**
Zhejiang Fortune Financial Center, Hangzhou, China, **196**

Parks
Binns Park, Lancaster, PA, **93**
Bridge Street Town Centre, McKinney, TX, **224**
Columbia River Park, Columbia, PA, **92**
Ford's Landing, Alexandria, VA, **270**
Lane Field, San Diego, CA, **132**
Nationwide Arena District, Columbus, OH, **190**
New Jersey Urban Parks Master Plan Competition, Trenton, NJ, **248**
North Bank Park, Columbus, OH, **189**
Schenley Plaza, Pittsburgh, PA, **246**
Shenzhen Bay Seafront Urban Design, Shenzhen, China, **150**
SouthPark Mall, Charlotte, NC, **146**

Plazas/Squares
Binns Park, Lancaster, PA, **93**
Culinary Institute of America Anton Plaza, Hyde Park, NY, **94**
Discovery Communications Headquarters, Silver Spring, MD, **252**
Downtown Silver Spring, Silver Spring, MD, **149**
Melrose Triangle, West Hollywood, CA, **223**
Sonoma Mountain Village, Rehnert Park, CA, **170**

WRIT Rosslyn Center, Arlington, VA, **269**
Zhejiang Fortune Financial Center, Hangzhou, China, **196**

Remodeling/Re-use
(including specifics on re-use of existing buildings; other re-used buildings may occur in Urban Redevelopment projects)
Cook County Circuit Courthouse, Chicago, IL, **58**
The Ferry Building Marketplace, San Francisco, CA, **42**
Hard Rock Hotel, Chicago, IL, **166**
Odeon Union Square, San Francisco, CA, **174**
Palmolive Building, Chicago, IL, **62**
The Robert Redford Building for the Natural Resources Defense Council (NRDC), Santa Monica, CA, **178**
Workhouse Art Center at Lorton, Lorton, VA, **266**

Residential Developments
10th and Hoyt Apartments, Portland, OR, **22**
16th/Market Affordable Residential Project, San Diego, CA, **130**
840 North Lake Shore Drive, Chicago, IL, **164**
Alexan Pacific Grove, Orange, CA, **40**
Amerige Pointe, Fullerton, CA, **208**
Aqua Vista Lofts, Fort Lauderdale, FL, **226**
The Architect Collection at Stapleton, Denver, CO, **70**
Arterra, Reno, NV, **24**
Aurora Condominiums, Silver Spring, MD, **168**
Avalon Del Rey, Los Angeles, CA, **204**
Beacon Point at Liberty Station, San Diego, CA, **258**
Beijing Century City East, Beijing, China, **30**
Belmar Row Houses, Lakewood, CO, **68**
Bridgeport Condominiums, Portland, OR, **20**
Bridges at Escala, San Diego, CA, **261**
Broadway Arms, Anaheim, CA, **172**
Broadway Grand, Oakland, CA, **173**
Chatham Square, Alexandria, VA, **160**
Cityville Fitzhugh, Dallas, TX, **128**
City Vista, Washington, DC, **274**
Clarendon Park, Arlington, VA, **156**
The Commons at Atlantic Station, Atlanta, GA, **124**
Deep Blue Plaza, Hangzhou, China, **134**
Del Mar Station, Pasadena, CA, **182**
Elizabeth Lofts, Portland, OR, **22**
Elleven, Los Angeles, **24**
The Garden, Taichung, Taiwan, **222**
Gregory Lofts, Portland, OR, **22**
Lake Bluff Tower, Milwaukee, WI, **26**
The Lofts at Stapleton, Denver, CO, **71**
Luma, Los Angeles, **24**
Mercer Square, Dallas, TX, **53**
Midtown Reston Town Center, Reston, VA, **154**
Mirabella, Seattle, WA, **23**
Odeon Union Square, San Francisco, CA, **174**
One South Market, San Jose, CA, **24**
Park Kingsbury, Chicago, IL, **168**
The Pinnacle Lofts, Portland, OR, **20**
Ponto Beachfront Village, Carlsbad, CA, **262**
Portico, San Diego, CA, **260**

Port Imperial, West New York, NJ, **158**
The Residences at the Greene, Beaverton, OH, **276**
Rollin Street, Seattle, WA, **23**
Shenzhen Bay Seafront Urban Design, Shenzhen, China, **150**
Sitka Apartments, Portland, OR, **22**
SouthPark Mall, Charlotte, NC, **146**
Tanner Place, Portland, OR, **20**
The Towne of Seahaven, Panama City Beach, FL, **148**
The Townlofts at Stapleton, Denver, CO, **71**
Tuck-Under Avenue Row Houses at Stapleton, Denver, CO, **72**
Vantaggio Baldwin Hills, Los Angeles, CA, **280**
Watermarke, Irvine, CA, **202**
West Highlands, Atlanta, GA, **126**
WRIT Rosslyn Center, Arlington, VA, **269**
X/O, 1712 South Prairie Avenue, Chicago, IL, **162**

Retail/Shopping Centers
(does not include street-level retail as part of primarily office or residential facilities)
Bella Terra, Huntington Beach, CA, **221**
Boca Raton Mixed-use, Boca Raton, FL, **80**
City Crossing, Shenzhen, China, **240**
The Ferry Building Marketplace, San Francisco, CA, **42**
Galerias Hipodromo, Tijuana, Mexico, **219**
Ibn Battuta Mall Phase II, Dubai, UAE, **78**
Oxbow Public Market, Napa, CA, **46**
Palladium, Birmingham, MI, **142**
Plaza Norte, Santiago, Chile, **286**
Riverside Plaza, Riverside, CA, **34**
Seattle Premium Outlet Center, Tulalip, WA, **38**
The Shops at Highland Village, Highland Village, TX, **214**
Short Pump Town Center, Richmond, Va, **286**
Somerset Collection South Renovations, Troy, MI, **140**
SouthPark Mall, Charlotte, NC, **146**
Sugarland Town Center Expansion, Sugarland, TX, **216**
Technology Square, Georgia Institute of Technology, Atlanta, GA, **284**
Town Center of Virginia Beach, Virginia Beach, VA, **82**
Towson Town Center, Towson, MD, **74**
Triangle Town Center, Raleigh, NC, **287**
The Village of Rochester Hills, Rochester Hills, MI, **138**

Sports/Recreation Facilities
Arena District, Columbus, OH, **190**
Downtown Detroit YMCA, Detroit, MI, **250**
Nationwide Arena District, Columbus, OH, **190**

Streetscape Improvements
DART CBD Transit Mall, Dallas, TX, **242**
Downtown Silver Spring, Silver Spring, MD, **149**
Ford's Landing, Alexandria, VA, **270**
WRIT Rosslyn Center, Arlington, VA, **269**

Transportation Facilities
DART CBD Transit Mall, Dallas, TX, **242**
Gateway Center, Los Angeles, CA, **110**

Urban Redevelopment
(including re-use of existing construction and infrastructure; see also Communities/Neighborhoods)
3949 Lindell Boulevard, St. Louis, MO, **52**
Arena District, Columbus, OH, **190**
Boca Raton Mixed-use, Boca Raton, FL, **80**
Chatham Square, Alexandria, VA, **160**
City Center at Oyster Point, Newport News, VA, **86**
Cityville Fitzhugh, Dallas, TX, **128**
Cityville Greenville, Dallas, TX, **122**
Cityville Southwestern Medical District, Dallas, TX, **125**
Clarendon Park, Arlington, VA, **156**
Columbus Riverfront Office Building, Columbus, GA, **118**
The Commons at Atlantic Station, Atlanta, GA, **124**
Cooper's Crossing, Camden, NJ, **276**
Court Street West Specific Plan, San Bernardino, CA, **218**
Crocker Park, Westlake, OH, **186**
Del Mar Station, Pasadena, CA, **182**
Downtown Long Beach Visioning, Long Beach, CA, **220**
Downtown Upland Master Plan, Upland, CA, **280**
East Fraserlands, British Columbia, Canada, **102**
Eastside, Richardson, TX, **51**
Ford's Landing, Alexandria, VA, **270**
Gateway Center, Los Angeles, CA, **110**
Hollywood & Highland, Los Angeles, CA, **112**
Lane Field, San Diego, CA, **132**
Melrose Triangle, West Hollywood, CA, **223**
Museum Place, Fort Worth, TX, **127**
Nationwide Arena District, Columbus, OH, **190**
Paxton Walk, Las Vegas, NV, **206**
Providence 2020, Providence, RI, **244**
Rio Nuevo, Tucson, AZ, **184**
Rocketts Landing, Richmond, VA, **88**
Taylor Yards Redevelopment, Los Angeles, CA, **280**
Technology Square, Georgia Institute of Technology, Atlanta, GA, **284**
Town Center of Virginia Beach, Virginia Beach, VA, **82**
Treasure Island Master Plan 2005, San Francisco, CA, **48**
West Highlands, Atlanta, GA, **126**

Preface

Richard M. Rosan, FAIA
President,
Urban Land Institute Worldwide

The green revolution is upon us, climate change concerns are everywhere, and sustainability is now a principal issue in real estate businesses of all types. Today architects and real estate developers alike are embracing sustainable development and green building tools and techniques, and this book highlights many fine examples of projects that have successfully embraced these practices.

Urban Spaces No. 5 illustrates a wide variety of designs and project types that use sustainable and environmentally-friendly approaches and techniques. From urban mixed-use properties to streetscapes to suburban town centers, the urban spaces presented in this book represent some of the best thinking in urban and environmental design today.

The Urban Land Institute is pleased to be cooperating with the publishers of *Urban Spaces No. 5* in this effort; place making and the creation of great urban spaces is fundamental to the healthy growth and development of cities and suburbs alike. ULI has nearly 38,000 members in 90 countries around the world working to create better buildings, places, and communities. Since 1936, ULI has attracted leading thinkers, designers, developers, and real estate practitioners in land use, urban planning, design, finance, and real estate development.

With each edition, *Urban Spaces* showcases global innovation in urban design. The firms and projects in this book offer examples of best practices worldwide, and represent a rich source of ideas and innovative thinking about both good urban design and sustainable development.

Hollywood & Highland by Ehrenkrantz Eckstut & Kuhn Architects, aerial view of development.

Introduction

Green by Degrees: Buildings, Urban Spaces, Communities

by John Morris Dixon, FAIA

By now, even people with little interest in environmental issues have heard about "green buildings." An issue that mattered only to a few zealots a decade ago, the subject of green building has expanded globally and is now the focus of countless design professionals, corporate executives, and government authorities. Even real estate ads tout the green-ness of office buildings and condominiums, anticipating its appeal to tenants and buyers.

Several historical strands have converged to form the current green movement. There was the environmental

Seaside, Florida, by Duany Plater-Zyberk & Company. Dating from 1981, the plan of this resort community embodies principles that would later be identified with sustainable development and applied in more recent communities documented in this book, including several by Duany Plater-Zyberk.

Photo: Courtesy of DPZ.

The Robert Redford Building for the Natural Resources Defense Council (NRDC), Santa Monica, California, by Moule & Polyzoides. A small office building demonstrates cutting-edge environmental design, reworking an existing structure, using materials selected for sustainability, and generating no carbon dioxide emissions at all.

Photo: Tim Street Porter.

movement of the 1960s, when Rachel Carson warned of widespread natural destruction and regional planners led by Ian McHarg boosted the hitherto obscure word "ecology" to public consciousness. Energy conservation rose to international attention around the time of the petroleum market shocks of the 1970s, spawning lots of energy-conscious building design that now rates a second look.

Each of these movements subsided all too quickly, but today's green movement seems to be here to stay. It combines a concern with the natural world and ecological sensitivity with responses to two overriding present and future concerns: global warming and the political/economic fallout of our abject reliance on petroleum. With the current consensus that carbon compounds produced by combustion are behind accelerating climate change, there is strong, justifiable pressure to reduce this combustion, with zero carbon emissions as the ideal goal. And, as nations continue to contend for the world's remaining petroleum sources, it is clear that the whole world's addiction to oil carries an immense economic and political cost.

"Sustainability" is another current term that more or less equates with "green" and is used roughly as often, but springs from a slightly different, if related, set of concerns. A sustainable environment is one that will not deplete irreplaceable resources – whether petroleum or mineral ores or forests. Sustainability also stresses saving or salvaging existing buildings and other artifacts, so that the material and energy that went into them is not squandered.

LEED Standards for Green Design

Along with the spread of green-consciousness has come wide use of the acronym LEED – for Leadership in Energy and Environmental Design. LEED certification, in its vividly termed gradations of Silver, Gold, and Platinum, is much sought after and widely boasted about by building owners.

10th and Hoyt Apartments, Portland, Oregon, by Ankrom Moisan Architects. One of 14 projects by the firm contributing to the mixed-use, mixed-income revival of the city's Pearl District, this building features a green roof over parking below, with downspouts channeling rainwater through detention basins into a cistern.

Photo: Koch Landscape Architects.

X|O, 1712 South Prairie Avenue, Chicago, Illinois, by Lucien Lagrange Architects. Supporting neighborhood revival with two apartment towers, plus townhouses and retail at street level, this development has been designed for LEED certification, with proximity to public transportation, a small on-site park, and a planted roof over its garage.

Promulgated by the U.S. Green Building Council, LEED certification is primarily an American standard, with different green-rating systems prevailing in some other countries.

There is every reason to celebrate individual green buildings – structures that require significantly reduced energy for heating, cooling, ventilation, and lighting, that make maximum use of recycled and recyclable materials, that have minimal construction waste, that meet high standards of indoor environmental quality.

Tapping non-polluting sources of energy, such as geothermal, is obviously worthy of LEED points, as is re-use of rainwater and waste water. The use of materials produced nearby can reduce energy consumed in transporting them, and factory production of major parts of a building can sharply reduce waste of materials. (It has been reported that at a typical construction site, 30 percent of the material ends up at the dump.) Re-use of existing construction and building components is also praiseworthy and earns LEED credits.

It is especially hard to quantify the value of a building's shape and orientation, the factors most attributable to the architects, although their effect is reflected in its measurable energy, lighting, and ventilation requirements. Some LEED points for buildings involve their location, with credits granted for brownfield redevelopment and for access to public transportation. Certain locations actually rule out obtaining LEED certification: sites on prime agricultural land, for instance, on flood plains, or where construction threatens wetlands or endangered species.

The design of the site around the building is also significant, with points awarded for providing open space, efficient use of water resources, reducing the heat-island effect, reducing light pollution, and managing storm water runoff. Individual projects are also rewarded for steps to reduce the volume of private vehicles, such as offering preferential parking for van pools and providing bicycle commuters with racks, dressing rooms, and showers.

Cook County Circuit Courthouse, Chicago, Illinois, by Booth Hansen. Adapted from an old warehouse, the building embodies several green strategies to qualify it for LEED Silver.

Photo: Mark Ballogg

China Science and Technology Museum, Beijing, China, by annex|5. In this structure to be located in the 2008 Olympics Park, most of the building volume will be underground to save heating and cooling energy, and aerated garden pools will serve as heat sinks for the air-conditioning system.

Hard-to-Measure Qualities

A concern that is hard to quantify is adaptability–sometimes more vividly called "loose fit." If structures and neighborhoods can eventually be adapted to serve purposes different from their original ones, great reductions can be saved in construction materials and labor, as well as disposal of demolition products. Consider New York's SoHo district and similar areas in other cities, where old industrial lofts have been adapted for residential and arts uses.

Advances in computers have clearly helped. Now it is possible much more readily than before to determine what structural configuration will yield the minimum demand for materials. And the energy effect of numerous internal arrangements and mechanical system designs can be computer tested to establish clearly how energy savings can be accomplished.

One negative lesson promulgated by many of the buildings from the 1970s period of energy-conscious design is that so much of it was accomplished at a sacrifice of aesthetics. Too many of the adventurous and instructive buildings of the time looked like giant mechanism that would not fit well into their communities. This was in the period, some may recall, when President Jimmy Carter set the example by wearing cardigans to allow his thermostat to be set lower, conveying the notion that energy saving called for sacrifices.

By contrast, architects of many of today's most sustainable buildings testify that making their buildings greener has actually enriched their designs and made them more complementary to existing and planned community development. Advances in materials have helped to make this possible: glass that admits light but rebuffs heat rays, for instance, no longer has to look mirrored or menacingly dark. Planted rooftops – usually aesthetic assets – are used more often now, and with greater confidence, than in the 1970s, in part because of more dependable materials and advances in their construction and planting techniques. In any case, it's obvious that the added constraints of green design produce better architecture only with the kind of skill and commitment

New Jiang Wan Cultural Center, Shanghai, China, by RTKL. Developed on Shanghai's last wetland preserve, the structure is designed to merge with its site and demonstrates many green methods and materials, including a composite wood cladding that is new to China.

Photo: Fu Xing Studios.

Visteon Village Corporate Headquarters, Van Buren Township, Michigan, by SmithGroup. These offices form a village laid out along a lake created from a disused gravel quarry.

Photo: Laszlo Regos Photography.

that has always produced superior design.

Lurking behind any discussion of green building is the question of cost. More effective building envelopes and mechanical systems often – but not always– involve additional first costs. While actual reductions in quantities of materials usually save money, cost savings in the reuse of old buildings and the recycling of materials may turn out to be largely theoretical. Reductions in the demand for energy and water will always result in lower long-term costs, but not initial savings. So generally speaking, designing greener buildings is likely to mean spending somewhat more on construction to reduce operating expenses down the road. Considerations of the building's life-cycle costs are typically the strongest economic arguments for building green.

Other management consideration may involve the pride owners, tenants, and occupants take in the building. Owning or inhabiting a building with exemplary environmental characteristics has a value in the marketplace; it can help recruit and retain employees or to sell condos or concert subscriptions.

Like exceptional aesthetic qualities, sustainability now has strong subjective value – a relatively new value in our society.

Beyond the Individual Building

The environmental gain in constructing even the greenest building, however, means little if it can only be reached by private vehicles. (Alternative fuels would help, but we're at least decades away from zero-carbon ones.) To really accomplish a green or sustainable environment, we need to think and plan on a community scale – even a regional scale – as well. We need to plan so that the public can circulate on foot or by shared circulation, rather than relying on private vehicles. We need to plan for full utilization of public utilities and public transportation facilities that already exist – retaining the investment already made in them. And we must

Technology Square, Georgia Institute of Technology, Atlanta, Georgia, by TVS. Connecting Georgia Tech with midtown Atlanta, this multi-block mixed-use development includes the College of Management, the 14th building in the nation to earn a LEED Silver rating for its environmental features.

Photo: Brian Gassel/TVS.

New Jersey Urban Parks Master Plan Competition, Trenton, New Jersey, by Sasaki Associates, Inc. One of five finalists in a master plan competition for the state capital's parks, this scheme would restore natural systems along the Delaware River and provide for recreation and environmental education.

do our best to ensure that our immense investment in existing buildings is not thrown away.

Recognizing this need, the U.S. Green Building Council has been developing a rating system for "Neighborhood Development." A pilot version of this system was made public in February of 2007 and will be applied to some control projects to test its effectiveness as a measure of sustainability. It is the latest in a growing series of LEED rating documents. Starting with the granddaddy LEED for New Construction and Major Renovation – latest version issued October 2005 – they include versions tailored for core-and-shell buildings, commercial interiors, schools, and others – some of them still in draft form.

The Neighborhood LEED rating program has taken longer to develop than some others because the measurement of sustainability for whole communities is inherently more complicated than for single buildings. It is conceptually easier to judge reductions in energy demand or storm water runoff – whether for one building or a whole community – than it is to decide what reduction in private car use can be attributed to any neighborhood or urban plan. And how can an inner-city infill development be compared environmentally with a planned neighborhood on the urban fringe?

Reflecting the complexity of judging some desirable "neighborhood" characteristics, the pilot rating system allows for much latitude in interpretation. Out of 106 possible total points, it allows 2 to 10 points for "reduced automobile dependence," perhaps the least predictable outcome, and 1 to 7 points for "compact development," a characteristic that could obviously vary widely depending on the situation.

Other items on the checklist that can earn more than one point include: brownfield redevelopment, proximity to housing and jobs, diversity of uses, inclusion of affordable rental housing, reduced parking footprint, walkable streets, transportation demand management, storm water management, plus up to 12 points for the sustainability characteristics of the buildings included in the neighborhood. LEED certification under this program requires a minimum of 40 points, with LEED Silver 50, Gold 60, and Platinum 80.

Culinary Institute of America Anton Plaza, Hyde Park, New York, by Derck & Edson Associates. Parking for an existing complex has been provided in a garage notched into the steep site, topped with a formal landscape that provides required filtering for storm water entering the Hudson River.

Photo: Nathan Cox Photography.

Salishan, Tacoma, Washington, by Torti Gallas and Partners. A mixed-use program transformed this former public housing site on a sensitive watershed into a 1,180-unit neighborhood sustainable in all its aspects – social, economic, environmental, and cultural.

Photo: Steve Hall @ Hedrich Blessing

As Green Design Evolves

While many projects in this book are being considered for LEED certification, there are good reasons why few of them have earned it yet. Even at the scale of the single building, where LEED standards are well-established, buildings have to be completed before the certification process can take place. And the outcome will depend in part on provisions for maintenance that doesn't jeopardize their green characteristics.

For multi-building complexes and neighborhoods, LEED standards have not yet been firmly established. And for complete new communities, they may be somewhere in the future. Many of the works cited in this book are, in effect, setting the standards for future LEED programs. Increasingly, sustainability standards are being required for construction projects by government organizations, such as the General Services Administration, the federal government largest builder, and for the projects built by or partially supported by many national, state, and municipal agencies.

As this is written, the U.S. Congress is considering a Carbon-Neutral Act, which would mandate staged reductions of the use of greenhouse gases for Federal projects, with the goal of zero carbon-based emissions by 2050. Many corporations and institutions are also mandating green design. And we are now beginning to see sustainability factors incorporated into municipal building codes that control all constuction within their borders.

Schools of architecture and urban design are also, slowly and somewhat belatedly, adding green design education to their programs. Faculty and students who have been largely immersed in considerations of form and expression will not be converted overnight.

But most schools of design are now recognizing that one required course in mechanical engineering is not adequate to cover the complex network of concerns behind green architecture and urban design.

Meanwhile, almost every project in this book makes a contribution to sustainability. They pick up the themes

The Commons at Atlantic Station, Atlanta, Georgia, by James, Harwick + Partners. At the heart of 138-acre Atlantic Commons brownfields redevelopment, this project is laid out around a five-acre park and features sustainable strategies, including separate waste and storm water systems.

Photo: Rion Rizzo.

Ford's Landing, Alexandria, Virginia, by Studio 39 Landscape Architecture. For this eight-acre townhouse development on a former industrial site, landscape strategies included crushing and recycling concrete paving, preserving adjoining wetlands, and providing an appealing pedestrian network.

that have run through all the volumes of the Urban Spaces series: the efficient use of land, the mixing of uses to create walkable environments, the design of pedestrian-friendly streets and plazas, reduction in the need to use cars and in the space required to park them. These are also objectives advocated by the Urban Land Institute, which has cooperated with Visual Reference Publications in the entire series of Urban Spaces books.

Sustainability can also be applied to whole cultures. In some societies, for instance, people tend to add or shed layers of clothing to compensate for differences in indoor temperature, rather than insisting on the same year-round temperature. Even different cooking methods can affect use of energy. And sustainability goals can be applied to whole communities, considering ways to keep them occupied and economically viable.

Narrowly viewed, green or sustainable design is seen as primarily a technical issue. And it does involve serious technical aspects of design, construction, and operation. But the sustainability of whole communities depends ultimately on all of the political, economic, and cultural factors that keep them and their nations healthy.

North Bank Park, Columbus, Ohio, by MSI. Recycling of structural and paving materials was featured in the transformation of 14 acres of brownfields adjoining downtown Columbus into a riverfront park adjoining the 79-acre mixed-use Arena redevelopment planned by the same firm.

Photo: MSI, Feinknopf.

Downtown Long Beach Visioning, Long Beach, California, by Perkowitz+Ruth Architects/Studio One Eleven. For a 150-block area under development pressure, this study proposes zoning that recognizes solar access and view corridors, encourages public transportation, and minimizes surface parking.

Ankrom Moisan Architects

Portland
6720 SW Macadam Avenue
Portland, OR 97219
503.245.7100
503.245.7710 (Fax)
www.amaa.com

Seattle
117 S Main Street
Seattle, WA 98104
206.576.1600
206.447.5514 (Fax)

Ankrom Moisan Architects

The Pearl District
Portland, Oregon

1996 — 1997 — 1998 — 1999 — 2001 — 2003

❶ ❹ ❺ ❼ ❽ ❾ ⑫

The Pearl District, located just north of downtown Portland, is a vibrant urban neighborhood named by Project for Public Places as one of the World's Best Places. Ankrom Moisan has played a dominant role in its development, designing 14 key projects in the Pearl, from the first conversion of a warehouse into loft housing to the most recent, larger-scale condominiums. Each project has contributed to the neighborhood's active pedestrian realm, with a diverse array of retail shops, galleries, and restaurants lining the streets. A range of housing options was a key city goal, and among the projects are affordable housing complexes, townhouses, live/work units, apartments and condominiums designed for a variety of economic strata. A series of parks provides public gathering spaces, and a streetcar line connects the Pearl to downtown and another live/work district to the west. Ankrom Moisan's architectural projects have been integral to the success of the parks and streetcars by providing active edges and enclosure for the parks and orienting building entries and retail to relate best to the streetcar. In Jameson Park in the heart of the district, for example, hundreds of people – including parents with their children – frolic in the fountain, play bocce ball or lounge and people watch. Three Ankrom Moisan projects surround and engage the park with street-level activity and housing above. These buildings also help to keep the park safe, with activity and eyes on the park nearly 24/7. Tanner Park to the north, designed by Atelier Dreistl, is framed by two Ankrom Moisan projects, the Pinnacle and Bridgeport condominiums. A restaurant pavilion in front of the Pinnacle acts as a beacon at night, activating the park. Restaurants and retail in the Bridgeport face the park, providing a bookend for the Pinnacle pavilion. The district includes several sustainable design strategies: locating housing downtown, where people can walk to work and are linked by public trans-

Top: Timeline for Ankrom Moisan projects in the Pearl District.

Opposite: Map and project key for Moisan projects in the Pearl District.

Photography: Jeff Krausse, Kirsten Force, Jeff Hamilton, Janis Miglavs, Michael Mathers.

2005 —— 2006

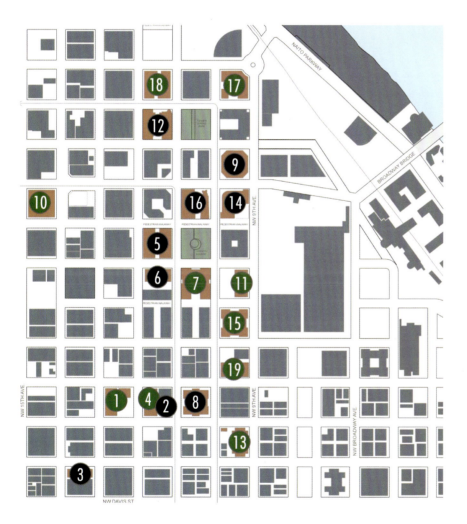

1. Chown Pella Lofts
2. 11th & Glisan Office Remodel
3. Bullseye Glass
4. McKenzie Lofts
5. Riverstone Condominiums
6. Johnson Street Townhomes (Interiors)
7. Tanner Place
8. Gregory - Mixed Use
9. Lovejoy Station Apartments
10. Marshall Wells Lofts
11. Pearl Clinic and Pharmacy (Interiors)
12. Bridgeport Condominiums
13. Elizabeth Lofts
14. Burlington Apartments
15. 10th & Hoyt Apartments
16. Park Place Condominiums (Interiors)
17. The Pinnacle Condominiums
18. Sitka Apartments
19. 937-Lofts

● = Sustainable Design Projects

Ankrom Moisan Architects

portation to the rest of the city and the region; adaptive reuse of existing buildings; environmentally sensitive treatment of storm water. An interior courtyard in the 10th and Hoyt Apartments, designed in association with Koch Landscape Architects, features a green roof over covered parking below. Three copper downspouts channel rainwater into detention basins and a cistern in the courtyard. The water is then recirculated across sculptural metal boxes pierced by glass buttons and illuminated by interior lights, creating a playful visual display and a

Above: Tanner Place framing Jameson Park.

Right and middle: Ground-floor restaurants in Bridgeport Condominiums.

Bottom right: Lobby of the Elizabeth Lofts.

Opposite: The Pinnacle Lofts at the northern gateway to the Pearl District.

Photography: Janis Miglavs, Jason Higbee, Kirsten Force.

Ankrom Moisan Architects

soothing sound for residents. At the Elizabeth Lofts, 100 percent of the storm water is collected and filtered on site. The project, which was designed to LEED standards, also recycled 90 percent of construction waste and is 20 percent more energy efficient than required by code. The Pinnacle Lofts has high-efficiency glazing and a green roof to filter storm water. The 937 Condos, designed for LEED Gold and under construction, will have a green roof and energy-efficient systems and glazing.

Top: Courtyard of 10th and Hoyt Apartments with rainwater filtering featured in its design.

Top right: The Elizabeth Lofts, designed for LEED Silver.

Above: The Gregory Lofts, a district icon.

Left: Affordable Sitka Apartments featuring numerous sustainable strategies.

Photography: Koch Landscape Architects, Kirsten Force, Jeff Krausse, Michael Mathers.

Ankrom Moisan Architects

Downtown Projects
Seattle, Washington

Top: Rollin Street, designed to LEED standards.

Bottom: Mirabella, a high-rise continuing-care retirement community.

Renderings: Ankrom Moisan Architects.

The firm has four residential projects underway in downtown Seattle, including a 42-story apartment building, high-rise senior housing, and a mixed-use condominium project. The materials and form of the condominium project, Rollin Street, refer to the site's past as well as the emerging high-tech residential and multiuse future. Ground-floor retail, with double-height glass around the entry, generates street-level interest and activity. Building proportions and façade treatment reinforce the character of the street and neighborhood. All parking is underground. The project is on track to be LEED certified. A new model of senior housing, Mirabella will provide all levels of senior care, including independent living, assisted living, and health care services integrated vertically. The U-shaped building, which surrounds a secure courtyard, is 12 stories tall and contains 400 living units. The dining room is located on the 10th floor, with unobstructed views of Lake Union. Other common facilities include a bar lounge, an auditorium, a wine tasting room, a library, and an exercise facility with a pool.

Ankrom Moisan Architects

Other urban projects
Los Angeles, Sacramento, San Jose, Reno

Ankrom Moisan is helping to revitalize downtown areas in several cities. Elleven and Luma are the first new condominium projects in downtown Los Angeles in 20 years. Standing side by side at 13 and 19 stories, they are both mixed-use structures with ground-floor retail, and both are on track for LEED Silver. In San Jose, One South Market and the Axis Condominiums will enhance an existing downtown business and entertainment district with round-the-clock retail and residential activity. Incorporating Sacramento's Central City Design Guidelines, the Fremont Mews and L Street Lofts will enrich the unique sense of place of that city's midtown through mixed-use development with careful site planning. Arterra, a 16-story mixed-use condominium building in downtown Reno, will provide a new anchor for the emerging cultural district near the art museum. The project is on track to achieve LEED Silver.

Top left: One South Market, which will add new vitality to downtown San Jose.

Top middle: Arterra, in downtown Reno, expected to earn LEED Silver.

Top right: Luma and Elleven in Los Angeles, on track for LEED Silver.

Left: Elleven, in downtown LA, is on track to be LEED Silver.

Renderings: Ankrom Moisan Architects.

annex|5
An Epstein Design Group

600 West Fulton
Chicago, IL 60661
312.454.9100
312.429.8175
312.559.1217 (Fax)
ametter@annex5.net

New York
Tel Aviv
Warsaw
Beijing
Shenzhen
Los Angeles

annex|5
An Epstein Design Group

Lake Bluff Tower
Milwaukee, Wisconsin

This residential development is designed for a site directly north of downtown, on the edge of a bluff 50 feet above Lincoln Memorial Drive, which borders Lake Michigan. For this exceptional site, the proposal seeks to weave together land and water in an integrated landscape of gardens and structures. Toward this goal, three strategies were used: maintaining an open view to the lake from Prospect Avenue on the upper side of the site; integrating residential scale elements into the design; combining physical features of both lake and city. The most notable design strategy is the orientation of the tall slab on a diagonal across the site, so that it frames the view corridor to the lake from the city and permits the natural amenities of the shoreline to extend past the bluff line. The rotation of the slab also orients its single-loaded dwelling units directly toward the south, affording them optimum views of the lake, the city, and the Milwaukee Art Museum, with its distinctive new extension. At entry level, a wood deck surrounding a reflecting pool draws the characteristics of lake and lakefront into the development. Along the edge of the bluff, four townhouse units are elevated above the deck level to allow unobstructed views through to the lake.

Top left: Main slab and adjoining tower seen from east.

Top right: Project's key elements and site.

Right: Plan of site and buildings.

Opposite: Development seen from lakeside to south.

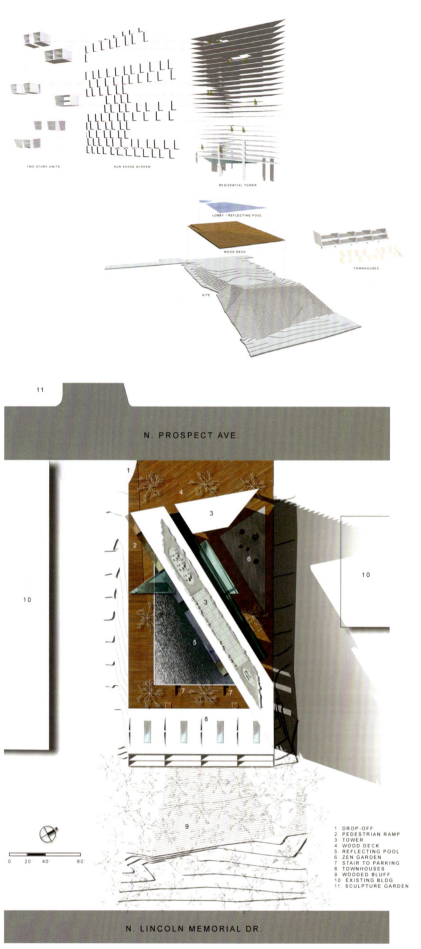

1. DROP-OFF
2. PEDESTRIAN RAMP
3. TOWER
4. WOOD DECK
5. REFLECTING POOL
6. ZEN GARDEN
7. STAIR TO PARKING
8. TOWNHOUSES
9. WOODED BLUFF
10. EXISTING BLDG
11. SCULPTURE GARDEN

annex|5
An Epstein Design Group

China Science and Technology Museum
Beijing, China

Taking advantage of its unique site geometry and prominent location within the 2008 Olympics Park, the museum has been composed as a set of discrete pavilions, each with a distinct function. These pavilions are set on a slightly raised green plaza that serves as a public park as well as a venue for museum events and outdoor exhibits. The majority of the museum spaces are located under this plaza. The entire composition is unified visually and functionally by the elevated children's museum exhibition hall, which floats above the plaza and follows the perimeter of the park, defining its boundaries. Symbolically, the integration of the green park with the museum serves as a powerful reminder that "nature" and "science and technology" are intimately related—technology derived from science and science from nature. The project has been designed to be sustainable and environmentally responsible. The bulk of it has been placed below grade, sheltered on all sides by earth, thus producing enormous savings in energy for both cooling and heating. The mechanical system has been designed to use the aerated garden pools as heat sinks for the air-conditioning system, and these pools will be replenished from rainwater storage tanks concealed under the landscaped berms.

Right: Museum seen from above, with elevated children's galleries defining perimeter.

Below: Low-level view, showing views into complex and supports for elevated perimeter galleries.

Bottom: Museum pavilions composed as discrete objects in a public park.

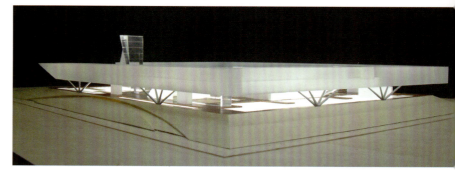

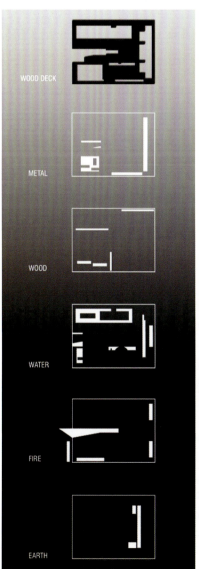

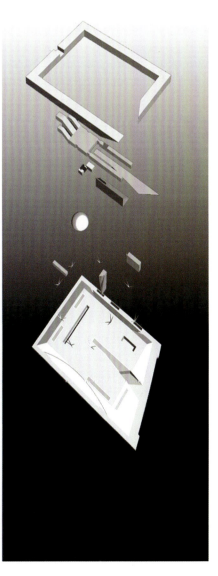

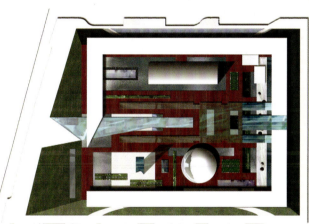

Left: Analytical drawings of museum's layers and components.

Above: Plan of museum on trapezoidal site.

annex|5
An Epstein Design Group

Beijing Century City East
Beijing, China

Left: Development after dark, with medium-rise apartment-hotel blocks at left, offset volumes of office tower at right.

Bottom left: Office tower by day.

Opposite top: Courtyard, with office tower to left, apartment-hotel building over commercial to right.

Opposite bottom: Stacking diagrams of project's functional components.

This project completes the final phase of Century City East, which is prominently situated at the intersection of the fourth ring road and the Beijing-Shengyang Expressway. On its site of 16,250 square meters (about 175,000 square feet), it proposes 96,500 square meters (about 1,039,000 square feet) of construction, including offices, commercial space, apartment-hotel, conference spaces, and support facilities. The client's goal was to create a signature office tower with maximized park views to the south, integrated with existing development. City requirements included a 100-meter (328-foot) height limit, a minimum of 30 percent green space, a maximum of 45 percent building coverage, and stringent solar access requirements that affected building placement. The office building's offset plan and sculpted form optimize the desired south views and exposure. The offset is defined by vertical atrium spaces at either end of the elevator core that allow through views. This feature also allows natural ventilation through convection. Dividing the apartment-hotel program into multiple volumes of varying heights created opportunities for spatial definition and view corridors. A maximum of four units per floor keeps the profile of these structures slender and contrasts to the standard hotel's anonymity. Instead of the usual retail podium, the commercial spaces meander through the complex to produce welcome pedestrian experiences and open the site to the public. The high-performance twin-shell building façades provide for natural ventilation, controlled daylight, and external shading. An integrated approach to internal systems allows users to control temperature, air quality, and illumination, while ensuring that unoccupied spaces are not maintained at maximum levels.

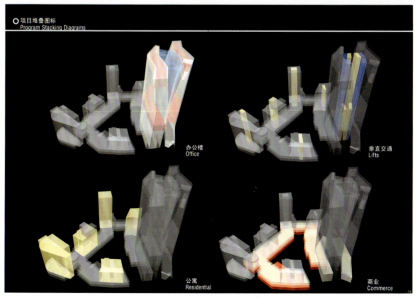

项目堆叠图标
Program Stacking Diagrams

办公楼 Office
垂直交通 Lifts
公寓 Residential
商业 Commerce

annex|5
An Epstein Design Group

Beijing CBD Office Park
Beijing, China

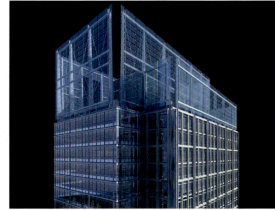

Located in the Beijing Central Business District among several high-profile developments, this 1.5-million-square-foot project includes three towers of Class A office space and a fourth tower providing condo-hotel services. At its base are three levels of retail, one of them below grade, and four levels of below-grade parking. The design applies advanced exterior cladding technologies to be provided by local manufacturers and installers. The firm worked with U.S.-based cladding consultants with offices in China to develop cladding that meets the design's aesthetic and technical requirements and can be delivered locally. Environmental features include high-performance low-e glass, exterior shading devices, operable windows, and individual climate control. The project also features landscaped green roofs over the retail podium and a unique glazed parapet that screens rooftop equipment and creates a distinctive skyline presence.

Above left: Study for entry canopy.

Above right: Study for parapet screen identifying complex on skyline.

Below: Four towers rising from project's retail podium.

Architects Orange

144 North Orange Street
Orange, CA 92866
714.639.9860
714.639.5286 (Fax)
architectsorange.com

Architects Orange

Riverside Plaza
Riverside, California

Riverside Plaza is a redevelopment of a 1960s enclosed mall on a 35-acre site. Executed in multiple phases, the 483,480-square-foot project includes: a fashion/lifestyle district anchored by the existing Gottschalks; a neighborhood district containing the expanded Vons grocery and the relocated SavOn Drugs and Trader Joe's; an entertainment district with a new 18-screen cinema, bookstore, restaurants, and shops along the new "Main Street." The distinctively curved "Main Street" features two large plazas, one at the food court and one at the theater entrance. The higher-end "Orchard Shops" at the opposite end of the complex share another public plaza, which serves as an outdoor dining space and a pedestrian link between the parking areas to the north and south. The project's complex phasing required construction of new buildings to relocate critical tenants before demolition of the mall could proceed. The redevelopment took six years to accomplish, with additional phases now being planned. At the request of the City Planning Department, the architectural style of the project was specifically not to be Spanish or Mediterranean. Instead, the contemporary design embodies Art Deco, Neo-Classical, Spanish Revival, and other influences, with the scale broken down so that the complex appears to be composed of separate buildings constructed over a period of time.

Opposite top: Before and after site plans.

Opposite bottom: Plaza in front of cinema along "Main Street."

Below: West gateway to "Main Street."

Bottom left: Bookstore at west gateway to "Main Street," with Gottschalks department store beyond.

Bottom right: Riverside Plaza's newest phase of shops at the west end of the project.

Photography: Barbara White.

Architects Orange

Piemonte at Ontario Center
Ontario, California

Below left: Rendering of typical street scene. **Below right:** Hotel pool deck with view of nearby mountains. **Bottom:** Buildings on plaza at main intersection.

Piemonte is designed as a mixed-use, 24-hour, pedestrian-oriented community near the intersection of Interstates 10 and 15 in a rapidly growing area of San Bernardino County. The 112-acre development will include 309,280 square feet of retail, 54,800 square feet of restaurants and services, 791 for-sale residences, 550,000 square feet of class-A office space, a 10,000-seat sports and entertainment arena, and a 311-room luxury hotel. The crossing of two key streets at the heart of the project will create a central plaza flanked by ground-floor restaurants with condominium units above. The east-west street is anchored by residential condominiums at one end and traditional retail at the other. At one end of the north-south street will be the arena and adjacent hotel. The architecture reflects the name Piemonte, a province of northern Italy and birthplace of a founder of Ontario. The large-scale buildings will be broken up visually to give the impression of structures built over time. Clay tile roofs, rough fieldstone, and simple volumes in traditional Italian colors – umbers, siennas, and ochres – will recall the original Piemonte. The landscape, as well, will suggest Italy, with olive, cypress, and pine trees along with grape vines. Environmentally, the construction materials and extensive shading devices will reduce cooling demands, the choice of plantings will reduce irrigation needs, and the density of the project will result in a more efficient infrastructure and lowered power demands.

Below: Plan of complete site, with retail in pink, restaurants in rose, residential in yellow, offices in blue, hotel in tan.
Bottom: Plan of mixed-use core.

SITE PLAN

MIXED USE DISTRICT
FLOORS 2-4

Architects Orange

Seattle Premium Outlet Center
Tulalip, Washington

Top right: One entry to central promenade.

Right: Outdoor dining under tensile fabric structure.

Opposite: Central promenade with canopies and awnings.

Opposite bottom left: Site plan, showing complex divided by midpoint promenade.

Opposite bottom right: Promenade after dark.

Photography: Milroy & McAleer, except top right photo, Larry Falke.

This factory outlet center contains approximately 120 tenant spaces distributed along an interior-oriented pedestrian promenade. At the center of the project, a public plaza provides the open portion of an indoor/outdoor food court. The interior promenade includes a series of interconnected "corner courts" and "mid courts." Cost-effective means were found to cover substantial areas of the interior promenade, in view of local weather conditions, yet admit as much natural light as possible. Construction cost was held down through use of tilt-up "building boxes," with independent tensile fabric structures covering the promenades. The fabric structures, trellises, and awnings that offer weather protection also control heat gain under sunny conditions. A desired "Northwest feel" in accented areas was created with independent single-sided structures finished with cedar slats, perforated and corrugated metal panels, and local river rock – materials that balance the natural and contemporary looks characteristic of the region. Colors derived from local flora, fauna, sky, and earth reinforce the regional feeling. Exterior paving is of concrete using exposed hard-seeded, water-washed glass aggregate in a variety of colors, with slate tile areas. Native plantings reinforce the regional character.

Architects Orange

Alexan Pacific Grove
Orange, California

Top right: Portion of façade with curved canopy at street corner extending along retail fronts.
Top: Whole project at night.
Above: Street frontages, showing variety of volumes, corbels, and cornices.
Left: Mid-block pool and recreation deck.
Photography: Steve Hinds, 2 upper photos; Tom Lamb, 2 lower photos.

Located in the "Uptown District" of Orange and replacing an unattractive motel, Alexan Pacific Grove is a mixed-use infill project providing 278 residential units and 5,000 square feet of retail space at a key intersection. The most challenging requirement from the developer was to "wrap" this mix of uses around a central parking garage in order to save on construction cost vs. underground parking. With the project's retail lying close to a major freeway exit, easy access to customer parking was required. The layout separates and facilitates parking for three types of users: shoppers, retail tenants, and residents. Resident amenities include a recreation area with fitness center and ample common and private open space. The architectural expression of the complex is contemporary, with articulated building volumes painted in bold colors, fifth-floor lofts activating its silhouette, and a curved metal canopy at the corner identifying the retail component.

BALDAUF CATTON VON ECKARTSBERG ARCHITECTS

1527 Stockton Street
4th Floor
San Francisco, CA 94133
415.398.6538
415.398.6521 (Fax)
info@bcvarch.com
www.bcvarch.com

BALDAUF CATTON VON ECKARTSBERG ARCHITECTS

The Ferry Building Marketplace
San Francisco, California

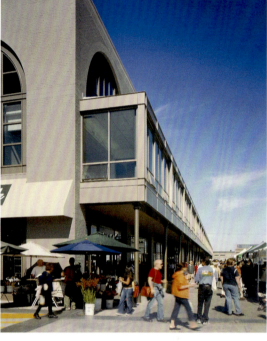

The rehabilitation of the landmark ferry terminal was a collaborative effort of BCV, as retail architects, with SMWM, architect, and Page and Turnbull, preservation architect. Major features of the project are the restoration of the Nave and Tower and the movement of the main public space from the second level to the street level, including removal of sections of the second floor plate to allow light to penetrate to the ground level. The ground floor of the interior has been transformed into a market

Top left: City side of Ferry Terminal, with vendors on plaza.

Top right: Remodeled Bay frontage of structure.

Left: Bay side of the Terminal, showing open air farmers market.

Opposite: Ferry Terminal Nave with Marketplace on ground floor, offices above.

Photography: Richard Barnes.

BALDAUF CATTON VON ECKARTSBERG ARCHITECTS

hall providing 65,000 square feet of restaurants, food producers, and purveyors. Combined with the outdoor farmer's market, the complex presents a unique celebration of San Francisco's regional bounty and multicultural culinary life. The upper floors include port offices and high-quality office space. In the grand-scaled public spaces, the historical and cultural significance of the building have been reinforced. Signage, displays, and restaurant design support the overall architectural concept and embody the traditions of the waterfront. About one third of the arcade shop area is reserved as incubator space, leased at below-market rates to foster entrepreneurs and encourage diversity of offerings. All of the individual tenant spaces shown on these pages have been designed by BCV.

Above: Capay Organic's produce shop along the Nave.
Bottom left: Taylor's Refresher restaurant featuring arched windows.
Below: Hog Island Oyster Co.'s bar and restaurant.
Bottom right: Ferry Plaza Seafood restaurant.

Above: Ferry Terminal, looking from downtown toward Bay.

Far left: Mijita restaurant.

Left: Ferry Plaza seafood restaurant.

Photography: David Wakely and Richard Barnes.

BALDAUF CATTON VON ECKARTSBERG ARCHITECTS

Oxbow Public Market
Napa, California

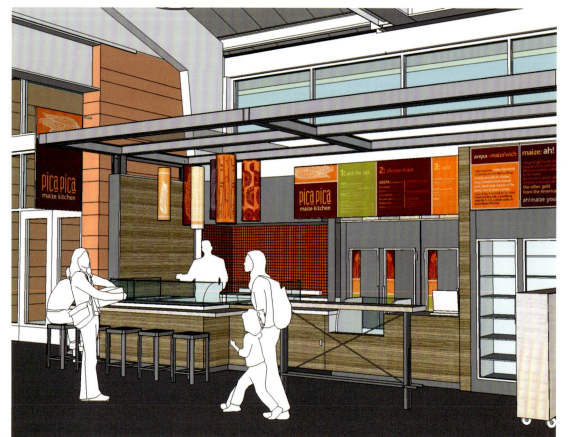

The proposed market is located in downtown Napa, adjacent to such food-related institutions as COPIA: The American Center for Wine, Food and the Arts; the Napa Valley Wine Train; the Napa Valley Farmers Market; and the Riverbend Resort and Spa. Containing food vendors and cafés, the market will showcase local organic farm products and artisan food and wine producers in an open-span public market environment. The market will adjoin the Napa River Trail now being built, and its wraparound promenade will encourage open-air shopping and community activity overlooking the river.

The Market Hall is a simple but elegant structure, taking cues from agricultural buildings. A prefabricated steel structure, it will have a low-pitched roof capped by a light and ventilation monitor along its ridge. The adjacent Wine Pavilion is conceived as a simple warehouse structure, with a cupola atop its hipped roof that admits daylight over the wine bar.

Top: Market Hall exterior, showing surrounding promenade.
Middle: River side entrance, showing the Main Hall and adjoining Wine Pavilion.
Bottom: Interior of Market Hall.
Illustrations: BCV Architects.

BALDAUF CATTON VON ECKARTSBERG ARCHITECTS

Mercato
Bend, Oregon

The project is located between downtown and Bend's old industrial area, now being revitalized as mills are converted into retail shops. Composed of several buildings, the Mercato will be a mixed-use center, with food retail, restaurants, offices, and residences. As design architect for the entire complex, BCV has incorporated a food marketplace with a double-height arcade, along the lines of their Ferry Building Marketplace in San Francisco. Open arcades and outdoor dining areas adjoin a path leading to the historic industrial buildings and the Deschutes River. Offices and condominium units above will afford spectacular views of the river and the National Forest beyond. Construction will be brick and concrete, with metal roofs and steel-framed windows recalling the area's industrial buildings.

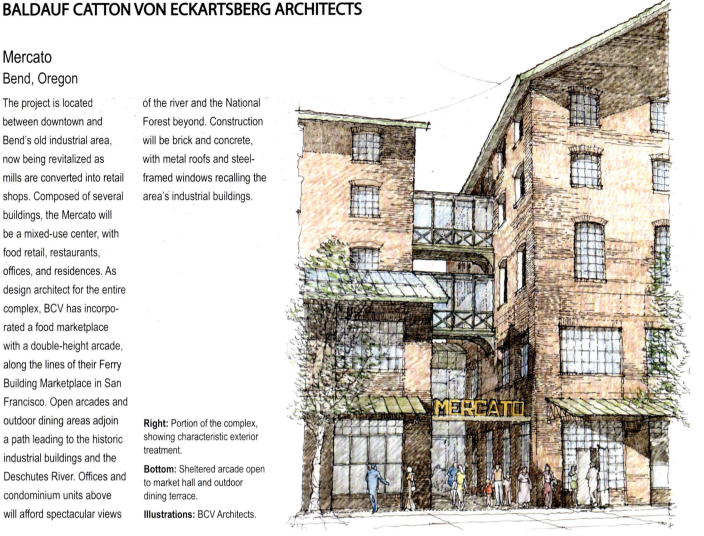

Right: Portion of the complex, showing characteristic exterior treatment.
Bottom: Sheltered arcade open to market hall and outdoor dining terrace.
Illustrations: BCV Architects.

BALDAUF CATTON VON ECKARTSBERG ARCHITECTS

Treasure Island Master Plan 2005
San Francisco, California

For this ambitious plan, BCV was responsible for retail design and related public spaces, working with Skidmore, Owings & Merrill (architecture and urban design) and SMWM (planning, urban design, and community development). The spectacular 403-acre site is at the very center of its metropolitan area, between San Francisco and the East Bay alongside the Bay Bridge. Treasure Island was man-made as the site of the 1939 Golden Gate International Exhibition. The plan proposes to create San Francisco's newest neighborhood, including housing, food retail, offices, condominiums, hotels, and a spa. The ecologically sensitive scheme reserves 60 percent of the island as open space, including 20 acres for an organic farm, while housing 5,000 people within a 10-minute walk of its ferry terminal, thus minimizing automobile traffic. Retail designed by BCV is concentrated in the dense hub. The plan also includes an expansion of the marina and recreational facilities, supported by sports and marine retailers.

Above: Model photo of Treasure Island's proposed high-density core

Left: Proposed development showing the reuse of historic hangar buildings and new residential towers.

Bottom left: Photomontage view of proposed island development from downtown San Francisco, with Ferry Terminal at left.

Illustrations: All images courtesy of Skidmore, Owings & Merrill.

Beeler Guest Owens Architects

4245 N. Central Expressway
Suite 300
Dallas, TX 75205
214.520.8878
214.520.8879 (Fax)
www.bgoarchitects.com

Beeler Guest Owens Architects

Eastside
Richardson, Texas

Construction started in October 2006 on a mixed-use community that will introduce a live-work-play milieu into the prime office corridor of Richardson. An existing office tower and a multistory parking structure will remain on the 14.8-acre site, and the completed development will include 435 residential units, 100,000 square feet of retail, 225,000 square feet of office space, and about 36,000 square feet of retail shops and restaurants – all laid out around a central 1-acre park with an interactive fountain and amphitheater. Residences will include a variety of loft and traditional-style units. New parking decks will be concealed behind one- and two-bedroom homes. While responding to the growing appeal of live-work-play communities, the project will also provide a retail, dining, and recreation focal point for neighboring corporate properties. Located just a few blocks from two Dallas Area Rapid Transit (DART) light rail stations, Eastside will offer a pedestrian-friendly environment to reduce reliance on the car. Architecturally, it will display sleek, contemporary forms coupled with the classic textures of brick and stucco. Occupancy of the residential units is expected in the fourth quarter of 2007.

Left: Distinctive architecture marking street intersection.

Middle left: Variety of architectural treatments for retail-residential structures.

Below left: Site plan, showing existing office and retail, with new residential, retail, parking structures, and extensive landscaping.

Bottom: Elevations along park frontage.

Opposite top: Apartments over retail bordering central park area.

Opposite bottom: Building elevations, showing architectural variety.

Beeler Guest Owens Architects 3949 Lindell Boulevard
St. Louis, Missouri

Top: Street frontage with residential over retail.   **Above:** Site plan.

In a location categorized as a "blighted area," 3949 Lindell Boulevard is a tax increment financed (TIF) redevelopment. It is a mixed-use, multifamily community being created, with the financial assistance of the city, to enhance public health, safety, and welfare. Developed in three clearly-defined phases, the 2.87-acre site will include 197 residential units, with 35,000 square feet of retail integrated into the mixed-use structures and 25,000 square feet of freestanding retail. The design of the project offers leeway for creative responses to prospective owners' needs, as well as market conditions as its completion progresses. The development is intended to alter the perception of the area as a residential destination and contribute to the neighborhood's stability by increasing property values and stimulating construction, permanent employment opportunities, and demand for services. More broadly, it is expected to demonstrate that St. Louis is a safe, affordable, and exciting place to live, increase the city's tax base, and encourage private investment in surrounding districts.

Beeler Guest Owens Architects

Mercer Square
Dallas, Texas

The Uptown district, just north of downtown Dallas, has become the focus of the city's contemporary culture and urban revitalization. Mercer Square will occupy one of Uptown's best remaining development sites, at the area's epicenter, next to the historic Greenwood Cemetery. Close to the district's shopping, entertainment, and arts destinations, the location has broad appeal for young professionals seeking residences to buy. For this site, the architects have designed a sleek, contemporary, four-story building featuring a curvilinear façade, expansive glass, balconies, and distinctive masonry and stucco details. Amenities will include a second-floor pool and deck for all residents and luxurious spa baths in individual units.

Top right: Site plan, showing residential units largely concealing parking structure.

Right: Treatment of prominent corner and long street frontage.

Bottom: Webster Avenue elevation.

Beeler Guest Owens Architects

Artessa at Quarry Village
San Antonio, Texas

Planned for completion in four phases, this development in the fashionable Lincoln Heights area will include 280 residential units averaging 1,142 square feet each. The complex will also include 35,000 square feet of retail integrated into mixed-use structures and 25,000 square feet of freestanding retail. In addition, a big-box retail structure is proposed for one end of the site and an office building for the far end. Parking will be partly on the surface and partly underground. Pools and other amenities on the site will be supplemented by the development's access to the adjoining Quarry Golf Course and Club.

Right: Mixed-use heart of development.

Below: Elevation showing transition from ground-floor retail to all residential.

Bottom: Overall elevation, showing retail frontages and variety of façade treatment.

Opposite bottom: Site plan showing phased development and adjacency to Quarry Golf Course.

Beeler Guest Owens Architects

The Triangle
Austin, Texas

Upon completion, The Triangle will include a total of 120,000 square feet of retail and restaurant space, much of it on the ground floor of buildings that will include 750 apartments. The development was controversial initially, since it replaces a hitherto park-like site that the community was accustomed to seeing as common open space, although it was not municipal property. Through more than a decade of public meetings and debates, this project evolved from one composed mostly of national retailers with large building footprints; to a mixed-use development, with retail shops more closely reflecting local needs. The complex is intended to illustrate the potential of innovative infill projects with a mix of uses and well matched in scale to surrounding neighborhoods.

Top and middle: Elevations showing variety of architectural treatment for project's mixed retail-residential buildings.
Left: Site plan.

Booth Hansen

333 South Des Plaines Street
Chicago, IL 60661
312.869.5000
312.869.5099 (Fax)
www.boothhansen.com

Booth Hansen

Circuit Court of Cook County
Chicago, Illinois

A former warehouse has been adapted as a $46-million, 177,000-square-foot courthouse to serve victims of domestic violence. Nine courtrooms have been located in the renovated 19th-Century building just south of downtown Chicago. To address the special needs of families involved, the building includes a dedicated and secure support area for victims, as well as a facility for the care, counseling, and education of children whose parents or guardians are attending court. To maximize use of the site and give the unique courthouse a civic presence, its previous rear elevation became its front. To reconcile this rear wall of "Chicago common" beige brick with the rust-colored brick of other walls, it was not simply reclad with matching brick. Instead, an atrium lobby was created by positioning a new wall 16 feet in front of the old one. The new wall applies a highly durable, minimum-maintenance tile system, using the rain screen principle, which is widely used in Europe but new to Chicago. Glazed openings, comprising 40 percent of this façade, lend the courthouse a stately yet welcoming appearance, and light the resulting 300-foot-long, 60-foot-high atrium. Wood panels on the inner surface of the new wall suffuse warm light entering through the south-facing clerestory at the top of the space. Used throughout the public interiors to create a restful, domestic feeling, the wood panels are resin-impregnated to resist scratches and vandalism. Oval raised ceiling areas provide dignified yet cheerful lighting and extend the building's civic spirit into the courtrooms and waiting areas. As the first facility carried out under Cook County's 2002 Green Building Ordinance, the courthouse was awarded Leadership in Energy and Environmental Design (LEED) Silver Level certification by the U.S. Green Building Council. Besides the rain-screen façade, sustainable strategies include proximity to public transportation options, photovoltaic roof panels supplying five percent of the building's annual energy, energy-efficient mechanical/electrical systems, and wood products certified by the Forest Stewardship Council. But the project's major green feature, especially considering the challenges of inserting thoroughly modern facilities into such a structure, is the reuse of the 19-Century building.

Top right: New entry façade built in front of original rear warehouse wall.

Right: Clerestory-topped atrium between new front and old wall.

Opposite middle left: Waiting area with oval ceiling recess.

Opposite bottom left: Typical courtroom.

Opposite bottom right: Corner view showing section of addition.

Photography: Mark Ballogg.

Booth Hansen

MB Financial Bank
Chicago, Illinois

A rapidly growing bank focused on serving middle-market businesses, MB Financial wanted an updated facility that would reflect its dynamic nature, reinforce its brand, and support a new marketing plan. Booth Hansen was commissioned to design a building to serve as the flagship West Loop branch. A study of the bank's requirements led to the design of a 50,000-square-foot, four-story steel and glass building with surface parking for about 20 cars. Integrated into the design is a prominent "lantern" at the top of the building, displaying the MB logo. Accommodating the bank's needs on the 11,800-square-foot site took ingenuity, especially since a basement on this property would have been cost-prohibitive. Functions normally relegated to a basement had to be divided between the ground floor and a rooftop penthouse. To maximize usable floor space, core and support functions are concentrated in a narrow precast concrete "box" extending from ground to roof along one side of the structure. The rigidity of this "box" eliminated the need for cross bracing in the steel framing of the open bank hall and the flexible office floors above it. The building envelope involves two distinct exterior wall treatments: an outer wrapping of green-tinted glass with stainless steel bands at the floor lines; and recessed street-level walls of clear low-iron glass supported by glass fins, stressing transparency around the main entrance. Inside this entrance, a floating stair of steel, stone, and glass reinforces the image of lightness and openness. Natural stone floors and wood paneling bring complementary warmth to the interiors.

Top left: Entire building at twilight.

Top right: One of two similar street fronts.

Above right: Entrance recess with bank hall beyond clear glass wall.

Right and opposite: "Floating" lobby stair inside clear glass-finned wall.

Photography: Barbara Karant.

Booth Hansen

Palmolive Building
Chicago, Illinois

One of the icons of Chicago architecture, the building was completed in 1929 to house the world-famous Colgate-Palmolive-Peet company. With its stripped-down Art Deco style and a luminous beacon added in 1930, the tower presided over Michigan Avenue as a symbol of the city's commercial strength. The 490,000-square-foot structure has now been transformed into a residential condominium building, one of the trailblazers of Chicago's commercial-to-residential conversion trend. Its $100-million renovation included design of 96 customized residential units, two floors of retail frontage, commercial offices, and a private club offering a health club and a walnut-paneled lounge to accommodate a range of entertainment events. The foremost challenge was providing parking on a site that had had none and

was landlocked, without alleyways. The solution was a pair of vehicular lifts to shuttle cars down to three levels of below-ground parking, construction of which required the removal of two grade-level columns and massive transfer girders to support the 37-story tower above. Antiquated mechanical-electrical-plumbing systems were removed while office tenants still occupied the building, and temporary measures were devised so new systems could be readily adapted for residential use. Raised floors on residential floors serve a dual purpose, accommodating mechanical systems and effectively lowering the height of window sills, which were too high for residential units. Mechanical risers were concealed behind the ornate doors of an unused elevator shaft, preserving the design of the building's elevator lobby. Crowning the renovation, the rooftop beacon was thoroughly restored and reinstalled.

Above left: Interior of private club.

Above: Tower in its prime location near Lake Michigan.

Left: Model residential unit.

Opposite: Walnut-paneled lobby.

Photography: Jon Miller, Hedrich Blessing; Tony Soluri (left); Scott Shigley (exterior).

Booth Hansen

The Joffrey Tower
Chicago, Illinois

State Street, Chicago's historic retail center, has been rejuvenated in recent years with new shops, theaters, restaurants, and residences. For a prominent site in the theater district at State and Randolph Streets, Booth Hansen has designed a 365,000-square-foot high-rise that includes luxury residential, retail, and a multi-function headquarters for The Joffrey Ballet of Chicago. The street-level retail is designed so that space on floors one and two can be occupied by a single tenant. The tower responds to an adjacent parking structure by being elevated above an open volume to give all 29 floors of condominiums spectacular views of Millennium Park and Lake Michigan. The ballet will unify all of its operations in a three-story, 35,000-square-foot permanent home. It includes six state-of-the-art rehearsal studios and a black box theater incorporating the latest dance technology, including floors that minimize distracting vibrations, so that artistic and administrative spaces can be intermixed without disturbing each other.

Left: Complex seen along State Street, framed by landmark Chicago Theater and Macy's (formerly Marshall Field's) department store.

Below: Lower portion of tower, seen from west along Randolph Street.

Braun + Yoshida Architects

B+Y Architects
1058 Delaware Street
Denver, CO 80204
303.623.0701
303.623.0560 (Fax)
www.braunyoshida.com

B+Y Architects

Belle Creek
Henderson, Colorado

Belle Creek is designed as a family-centered new community with all residential units within a five-minute walk of the town center. It was planned in collaboration with enlightened local leaders, who agreed with the developers on an increase in density, thus permitting a range of residential types – many of them more affordable than in comparable suburbs – and enabling the funding of generous community amenities. Of the 928 residential units, 185 are single-family houses of 1,581 to 2,331 square feet, 341 are detached "cottages" of 1,170 to 1,543 square feet, 98 are row houses of 1,019 to 1,599 square feet, and 304 are apartments. Residents are able to move to larger or smaller quarters without leaving the community. Site planning and architectural design were pursued simultaneously to create a coherent fabric. Residential architecture is inspired by established neighborhoods of Denver and Boulder. Garages are accessed via alleys, which are made safer and more attractive by bands of landscaping and the presence of carriage house units. Street and sidewalk design leaves ample room for mature trees. A 22,800-square-foot family center includes a fitness center, an early childhood learning center, and a technology center. A charter school built adjacent to the family center shares its gymnasium. A central 5,000-square-foot retail and professional office area includes a small grocery, florist and coffee shop. The 70,000-square-foot retail center at the south end of the project will grow as the local market develops. Measures to enhance sustainability include dual water systems, with nonpotable water for irrigation, and houses that meet "Built Green" energy-efficiency standards, which qualify owners for more advantageous mortgages.

Top left: Cottages facing park land.

Above left: Town Square with rowhouses and single family houses.

Above: Street lined with single family houses.

Right: Site plan.

Far right: 1½ story house with historically inspired porch.

Photography: Michael Peck, far right, Ron Ruscio.

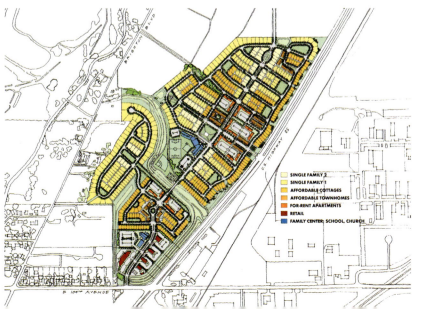

B+Y Architects

Belmar Row Houses
Lakewood, Colorado

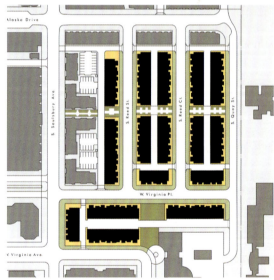

The new "downtown neighborhood" of Belmar is being developed, with residences, offices, shops, movies, galleries, restaurants, and plazas all within walking distance. The greyfield development is the site of a former regional mall. Mid-block walkways encourage pedestrian access to nearby business and entertainment destinations. B+Y has designed 132 row houses on 3½ blocks along the east side of Belmar. The three- and four-story residences include private second-floor decks, above the activity of the street, and optional penthouses offering more distant views. Gabled fronts refer to historical residential precedents and help to distinguish these buildings from the parapeted structures of adjacent retail blocks.

Top: Row houses with upper-floor decks and bays.

Above left: Residences along through-block walks.

Above: Site plan.

Photography: Cheryl Ungar.

B+Y Architects

The Brownstone Collection at Highlands Ranch Town Center
Highlands Ranch, Colorado

Top: "Stacked" brownstones facing town green.

Above: "Courtyard" brownstones facing the civic green.

Photography: Cheryl Ungar.

Once the epitome of urban sprawl, Highlands Ranch is now taking advantage of a unique opportunity to create a town center with civic, retail, office, and residential uses within walking distance. The first phase is 20.9 acres and includes 146 brownstones designed by B+Y Architects and 111 triplexes by others. Blending historical and contemporary architecture, the buildings bring urban imagery to a hitherto suburban community.

Elevations consciously define the public realm, while giving residents a semi-private envelope. Site planning and architecture were carried out simultaneously to solve grading, density, and placemaking goals. The grade was intentionally raised for houses facing the civic green and library in order to give the public space a clear boundary and to afford privacy for residents overlooking public events and celebrations.

B+Y Architects

The Architect Collection at Stapleton
Denver, Colorado

The development plan of Stapleton, on the site of Denver's previous airport, follows New Urbanist principles, and for the first several years, followed stipulations by the Stapleton Design Book to reflect architectural elements found in the city's older neighborhoods.

As development proceeded, the developer, builders and designers agreed to allow more divese and contemporary architecture. Street fronts of the Architect Collection houses face common greens, with sidewalks encouraging pedestrian activity, and garages are accessed from alleys. Three house types are identified by their defining elements: the Courtyard Residence by the outdoor room around which are wrapped living and sleeping areas; the Tower Residence by the third-floor loft above a two-story library; the Atrium Residence by a two-story, top-lighted living area and an optional two-story backyard pavilion.

Top left: Tower Residence with third-story loft above library.
Top right: Atrium Residence.
Bottom: Model houses at dusk.
Photography: Jim Blecha.

B+Y Architects

The Lofts at Stapleton
The Townlofts at Stapleton
Denver, Colorado

The Lofts are configured in a four-story crescent with an important design role: framing Stapleton's Founder's Green. The building is of Type V construction over an underground parking garage. The project's nine 1,450-square-foot lofts are two-story units with front entries on Founder's Green and private attached two-car garages. The 28 lofts on the third and fourth floors, ranging from 700 to 880 square feet, have stair and elevator access from the street-level lobby and the parking garage. Completing the rectilinear corner of the block are the Townlofts at Stapleton, which consist of 11 three-story row houses, varying in size from 1,060 to 1,489 square feet.

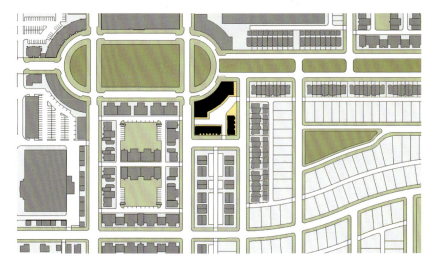

Top: Crescent-shaped building of The Lofts defining one quadrant of Founder's Green.

Above right: Detail of entrance.

Right: Site plan of The Lofts, with The Townlofts filling the rectangular portion of block.

Photography: Cheryl Ungar.

B+Y Architects

Tuck-Under Avenue Row Houses at Stapleton
Denver, Colorado

Like the developments on the previous two pages, this project contributes to the planned community of Stapleton, on the site of the old Denver airport. The Tuck-Under Avenue Row Houses line 29th Avenue, which forms the spine of Stapleton's first neighborhood. Varying in floor area from 1,385 to 1,609 square feet, the three-story houses hold the street edge in the manner of historical townhouses.

Above: Tuck-Under Avenue Row Houses defining key community street.
Left: Detail of Row Houses.
Photography: Cheryl Ungar.

Carter & Burgess, Inc.

Corporate Office
777 Main Street
Fort Worth, TX 76102
800.494.4082
www.c-b.com
design@c-b.com

Anchorage
Arlington
Atlanta
Austin
Baltimore
Boston
Chicago
Columbus
Dallas
Denver
Detroit

Fort Lauderdale
Houston
Inland Empire
Las Vegas
Little Rock
Los Angeles
New York
Oakland
Oklahoma City
Orlando
Phoenix

Raleigh
Sacramento
Salt Lake City
San Antonio
San Jose
Seattle
St. George
Santa Ana
Tampa
Washington, D.C.

Carter & Burgess, Inc.

Towson Town Center
Towson, Maryland

This project will transform a complex that has evolved over a period of 40 years, from an open-air retail center to an enclosed shopping center, to the status of a regional mall in the early 1990s. The current expansion and renovation provides an opportunity to redefine the relationship of the mall to its urban location – to reconnect it with the pedestrian on the street. A prominent new entrance from the street gives the project a "front door" for downtown pedestrians. Sidewalks and planting along the site's perimeter have been designed to meet requirements established for downtown Towson. The project's street frontage includes restaurants with outdoor seating. Building façades have been articulated to appear as a collection of related buildings, with bold environmental graphics indicating their collective identity. The redesign includes an updating of the main vehicular entrance, with textured paving materials, new landscaping and graphics, and a curved colonnade through which cars pass toward the garage, which has been expanded to 4,750 spaces. New retail area of about 81,000 square feet and restaurant area of about 28,000 square feet are being added, and the center's existing area of about 945,000 square feet is being remodeled.

Right: Towson interior sketch.
Below: Model view of new vehicular entrance.
Below middle: Overall model view.
Bottom: Elevation of new vehicular gateway and adjacent retail.
Opposite bottom: Street elevation.

Carter & Burgess, Inc.

Las Colinas Entertainment
Irving, Texas

Left: Entry plaza with fountain.
Below: Grand stair.
Bottom: Equestrian arena.
Opposite: Aerial view.

With a program modeled after the French Quarter of New Orleans, Las Colinas Entertainment is designed to be the premier entertainment destination in the Dallas area. The 40-acre project will house a concert hall, an outdoor equestrian arena, music venues, restaurants, a hotel, a civic center, retail shops, and two parking structures. The development's entry plaza is an adaptation of the Mexican zocalo, the public space that is the focal point of every community and the setting for its principal landmarks. In this project, the Civic Center, the Concert Hall, and the hotel will be the dominant presences on the focal plaza, which will feature a large fountain, a tall gateway, shrubbery, and flowers. Connectivity to the community will be established with a proposed extension of the monorail, which will have a station in the entry plaza. From this plaza, visitors can proceed to the pedestrian promenade Street of Music, lined with dining, entertainment, and shopping experiences. Architecturally, the Hispanic and Mediterranean heritage of the area's early settlers will be embodied in the Spanish Colonial Revival style. Detail, ornament, and massing are adapted from several eras of Spanish and Mexican architecture. The buildings will feature a combination of natural stone and smooth stucco wall finishes, terra cotta and cast-stone ornament, low-pitched clay tile roofs or flat roofs, balconies, round-arched windows and arcades, canvas awnings, and iron trim. The completed project, estimated to cost $600 million, will include a 6,500-seat concert hall, a 148,000-square-foot civic center, a 1,200-seat equestrian arena, a 33,000-square-foot VIP terrace, a 205,000-square-foot hotel, 190,000 square feet of entertainment venues (clubs, restaurants, retail), an 85,000-square-foot cinema, and 1,800,000 square feet of parking.

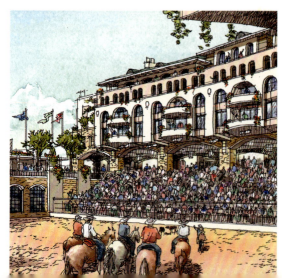

Carter & Burgess, Inc.

Ibn Battuta Mall Phase II
Dubai, United Arab Emirates

From the existing retail mall on this strategically central site, the high-rises that represent Dubai's booming economy can be seen in the distance. With the planned transformation of this complex into a mixed-use development, it will make its own contributions to the city's burgeoning skyline.

After an investment estimated at $2.4 billion, the project will include greatly expanded retail, along with dining, entertainment, living, working, hospitality and parking components, for a total of about 20,000,000 square feet. A new Arabian Souk will house gold retailers from around the globe, building on Middle Eastern traditions of quality and merchandising. Extreme sports ranging from snow-boarding to bungee jumping will be available for the young and young at heart. A walk through a glass tunnel will allow visitors to watch divers swim with sharks and sting rays. A science center will add another kind of adventure to the retail experience. A palm-lined boardwalk around a lagoon will link the attractions of the entertainment district. Many of the project's shoppers and businessmen will be able to commute home via a conveniently located metro station, while visitors will take elevators to their hotel accommodations or condominiums. Structured parking for patrons arriving by car will reduce the development's heat-island effect. The project's plans are notable for the way a broad boulevard, lined with an intense mix of uses, is superimposed on a lower level of retail expansion with a very different, but carefully coordinated, layout. The architecture of the expanded complex complements the themed design of the original mall, which paralleled the travels of a Renaissance-era Muslim explorer throughout the Islamic lands. In the interest of a cohesive development, the enlarged complex will continue this architectural learning experience with architecture reflecting the traditions of North Africa, Yemen, Arabia, Turkey, Southeast Asia, and Indonesia. The defining spirit of the new Ibn Battuta Mall will be the aim of a legitimate urban district which acknowledges the cultural history of the region, the technological invention of its current growth and its potential for representing new responsible and sustainable development.

Top left: India Court entrance to retail mall from podium level.
Above left: Entertainment district around lagoon.
Right: Podium Boulevard.
Opposite top: Podium-level plan.
Opposite middle: Ground-level plan.
Opposite bottom: Model of complex with gateway to boulevard in foreground.

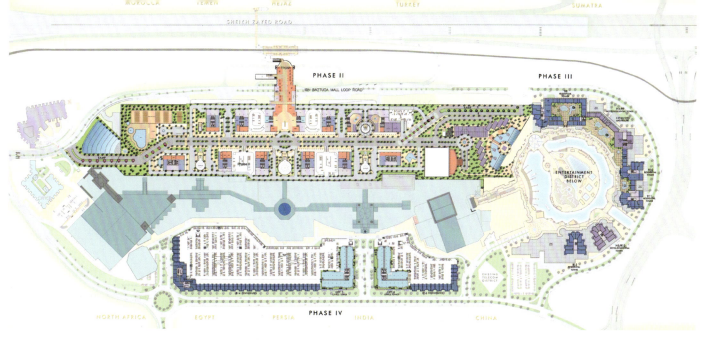
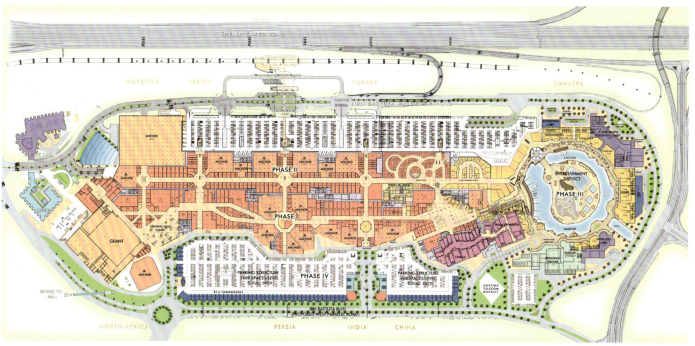

Carter & Burgess, Inc.

Boca Raton Mixed-use
Boca Raton, Florida

Capitalizing on its adjacency to a large regional mall, this transit-oriented development is planned to include 1,765,000 square feet of residential, retail, dining, entertainment, and office facilities on its 26 acres. While including a substantial amount of new construction, the project is designed to reflect the qualities of a small-town environment. Its architectural style will be both modern and indigenous to Florida. The streetscape will provide ample shade and protected outdoor seating. With ready access to Tri-rail, shuttle, and bus mass transit systems, the project should reduce automobile trips for its residents, workers, and visitors. By including a percentage of affordable work-force housing, it will promote social responsibility and reduce travel demands. The higher density and mix of uses will permit a lower ratio of parking spaces than if the uses were separated. The project's phasing will have to be carefully adjusted to create a viable mass of retail, allowing for the addition of residential above, without over-expenditure on construction in the first phase. Ultimately, the development is to include 1200 residential units, 100,000 square feet of offices, and 345,000 square feet of retail, restaurant, and entertainment.

Top: Urban Plaza along 19th Street.

Above right and right: Other streetscapes in development.

CMSS Architects, PC

4505 Columbus Street
Suite 100
Virginia Beach, VA 23462
757.222.2010
757.222.2022 (Fax)

www.cmssarchitects.com

11921 Freedom Drive
Suite 250
Reston, VA 20190

5000 Old Osborne Turnpike
Suite 200
Richmond, VA 23231

CMSS Architects, PC

Town Center of Virginia Beach
Virginia Beach, Virginia

The City of Virginia Beach, once an oceanfront resort town and agricultural community, has been facing expotential growth and dwindling land for new development. Now the largest city in Virginia, it has sought to direct growth, preserving its remaining farmland and addressing its lack of a true urban core. A group consisting of business leaders, a developer, and an architect, working with the city government, initiated planning and development of a $400-million mixed-use central business district. For this new core, the design team selected a 40-acre site for its proximity to the interstate, its central geographical position, and its location along an existing main commercial corridor. A "Main Street" style development was envisioned, with high-density construction on walkable blocks. The resulting master plan proposes a street grid enhanced with streetscape amenities, parks, and free public parking, with low-, mid-, and high-rise buildings for various uses. The grid also facilitates transitions to existing and potential new neighborhoods surrounding the center. The development will ultimately provide 1.8 million square feet of new construction over 17 blocks. Implementation in three phases began with ground-breaking in 2000. A city-approved Tax Increment Financing (TIF) district is funding public garages, land for a public plaza, and a future pedestrian bridge. Other city programs are supporting development of the plaza, a performing arts theater, and infrastructure improvements. The center will include more than 500,000 square feet of commercial/office space, 832,500 square feet of retail, and over 450 residential units. Phase One, now completed, includes 400,000 square feet of office space in a central 23-story tower, strikingly visible from the interstate, and a hotel. Streets around these structures are enlivened by shops, services, and restaurants. The second phase will include housing and arts and entertainment facilities, along with additional retail, restaurants, and offices. Phase Three will bring a vertically integrated high-rise that will be the tallest building in the state. Long before its completion, the development has established an identity for the city and begun a ripple effect of renovation and redevelopment in adjoining areas.

Top left: Retail uses around one of project's plazas.

Top middle: Mixed-use tower, visible from key highways, on central plaza.

Top right: Site plan.

Opposite bottom: Proposed arts center plaza and office building.

Photography: Steve Budman and CMSS Architects.

CMSS Architects, PC

Kincora
Loudoun County, Virginia

Left: Retail-lined street on outside of lagoon circle.

Below: Conceptual sketch displaying the potential volume and scale of mixed-use structures.

Bottom: Multiple uses, such as office and retail, integrated within the same block.

Right: Street-level retail with residential above around central lagoon, offices over retail on adjoining blocks.

Opposite bottom: Site plan with state highway 28 along east (bottom) edge and conservation area to west.

Developed in direct response to the tremendous growth in the Washington, DC metro area, Kincora is a 424-acre tract at a key highway intersection only five minutes from Dulles International Airport. 185 acres of watershed will be preserved as exemplary of the area's rural heritage. Sleek, high-profile buildings lining the property's east edge, along state route 28, will accommodate offices for the high-tech services sector and take advantage of the location's exceptional transportation connections. Toward the west, the scale of the development will step down into a pedestrian-friendly Village Center, with single-story restaurants and open-air markets bordering the natural park land. Amenity retail and service establishments will be within walking distance of offices. The development's high-tech office activities will be supported by hotel and conference center facilities, residences for students, researchers, and executives-in-training, and a resource center. A continuing-care retirement center intended for educators and researchers will facilitate further contributions to the development's highly educated work force. When completed, Kincora will total nearly 7.5 million square feet of mixed-use development. Walking and bike trails in the natural preservation area will be laid out to connect to a network of trails throughout the Northern Virginia region.

HOUSING		1,720,20
OFFICE 8 STORIES & UP		4,240,00
OFFICE 2 & 3 STORIES OVER RET.		558,00
RETAIL UNDER OFFICE		132,00
TWO STORY RETAIL / OFFICE		110,00
SINGLE STORY RETAIL		94,00
HOTEL		600,00
		7,454,20

CMSS Architects, PC

City Center at Oyster Point
Newport News, Virginia

Above: Conference center rotunda with attached hotel behind.

Top right: Mariners Row retail district.

Right: Buildings around central fountain lake.

Below right: City Center master plan.

Opposite top: Mixed-use development along lake.

Opposite bottom: The Point Condominiums.

Photography: Steve Budman, Len Rothman.

Stretching along the James River, the City of Newport News grew westward from the historical downtown at its east end, adjacent to its shipyards. Its commercial center has now shifted west to the area called Oyster Point, where the city owned a 52-acre parcel in the midst of commercial development. The subject of previous plans, it remained undeveloped. In 2001, a mixed-use development was proposed, incorporating new urban principles of stacked mixed uses in a walkable environment, with civic open spaces and waterfront parkland. The goal of this new plan was a town center including housing, offices, retail, hospitality and conference facilities, and outdoor public areas. Through extensive consensus building and public presentations, community support was generated, and the planning team worked closely with city officials on specialized zoning and standards that would encourage thoughtful development. The plan is based on a grid of walkable blocks, flexible in terms of uses, laid out so that projects frame public open spaces. A parking strategy supports the density required. When completed, the development will include one million square feet of office space in 12 buildings, 600 residential units, 250,000 square feet of retail, dining and entertainment, parking for 4,700 cars, and a 250-room hotel with conference center. Key public spaces are a five-acre fountain lake and an eight-acre park for festivals and recreation. The development has quickly become the social epicenter of the city.

CMSS Architects, PC

Rocketts Landing
Richmond, Virginia

For 400 years, Rocketts Landing has played a major role in development of Richmond, first as a center for international sea trade, then as the confederate shipyards, and finally as a major turn-of-the-centruy industrial center. Demolished in the 1970's as part of urban renewal, this 54-acre tract remained persistently resistant to development. Its brownfield status raised issues of remediation, liability, and cost. In 1999, the development team engaged the design team for a site analysis and feasibility study, to gain support from investors and the public. In 2002, the state granted advance approval for the remediation plan, the first large-scale multi-parcel project so approved in the Commonwealth. Division of the site between the City of Richmond and Henrico County, with neither jurisdiction allowing the intended mixed-use and density, required close cooperation with officials to craft new zoning ordinances. Success depended on extensive involvement with stakeholders on the site and in surrounding neighborhoods. The master plan envisions a 25-block community that can be adapted to current or emerging development trends, with a network of open spaces that connect the development to the James River. Existing natural sections of the riverfront are preserved, with paths and trails, and an urban waterfront created, with marinas, boat launch sites, and a mile-long public park. Components of the plan include: Festival Plaza, a public gateway from downtown Richmond; The Square, centrally located, with an urban mixed-use atmosphere; Village Commons, a gathering place with river views in an area where old warehouses are converted to office and residential use; East Village, a residential district with tree-lined streets leading to the riverfront. Construction of Phase One began in 2005 and is scheduled for completion in 2007.

Top left: Phase II mixed-use and riverwalk.
Top right: Aerial view of Phase I.
Middle left: Streetscape with preserved water tower.
Left: 210 Rock, condominiums above street-level retail.

Derck & Edson Associates

33 South Broad Street
Lititz, PA 17543
717.626.2054
717.626.0954 (Fax)
www.derckandedson.com

Derck & Edson Associates

Lititz Watch Technicum
Lititz, Pennsylvania

Left: Terrace retaining wall, pool, and pond, with vertical and horizontal jets.

Bottom left: Pond and buildings seen from direction of road.

Below: Technicum seen across pond at dusk.

Photography: Larry Lefever.

Derck & Edson Associates was involved in this project from the selection of the 4.1-acre site to the last detail of vehicular and pedestrian circulation, storm water management, and planting. The Technicum is a watchmaking/repair school, which includes classrooms, administrative offices, a library, and a cafeteria with an outdoor terrace. The location was selected for its small-town atmosphere and proximity to downtown destinations. The site has been designed to tie into the local vernacular and streetscape and provides contemplative spaces for students and employees during break times. Fountains and pools surrounding the dining terrace serve as sound buffers against nearby traffic. Stone is from local quarries, and terraces and walkways are paved with Pennsylvania bluestone. The pond does not serve, as might be expected, as storm water detention, which is handled entirely underground. Extensive plantings include a 10-inch-caliper beech at the pond and 4-inch-caliper honey locusts on the dining terrace. The building was designed by Michael Graves and Hammel Associates. Large north-facing windows provide bright natural light for the work spaces.

Derck & Edson Associates Columbia River Park
Columbia, Pennsylvania

Left: Storm water outlet into cove for canoe and kayak launching.

Below left: Central lawn and dining terrace.

Bottom left: Aerial view of central area.

Illustrations: Derck & Edson Associates.

The design for this 11-acre park along the Susquehanna River was developed with full community involvement and includes several key elements: upgraded pedestrian access, with new entry features; an accessible river walk with viewing platforms; a new marina building with a terrace for cultural programs and café dining; new and renovated boat launch facilities, including a canoe and kayak loading area; removable boat docks; interpretive biking and walking trails; an education center; an amphitheater; expanded parking areas. The program envisions an outlay of $7.5 million over a 7-to-10-year period. The outlet of a 60-inch storm drain will be reconfigured with a series of pool and ripple sequences to reduce sediment and debris entering the river. Invasive non-native plants such as bamboo and English ivy will be replaced with native species. Porous paving will reduce runoff into the river from the park. Buildings will be designed with the goal of LEED certification. An adjoining freight railroad line requires a secure fence, made into an attractive barrier with native grasses and small shrubs. The adjacent water supply facility also requires security barriers, which will be provided by a historically appropriate fence, which will also serve to celebrate the park entry.

Derck & Edson Associates

Binns Park
Lancaster, Pennsylvania

Right: Evening event in the park.

Bottom left: Central fountain in action.

Bottom right: Overall view of park and building embracing it, originally built for Armstrong World Industries, now county-owned and occupied by public and private offices.

Photography: Nathan Cox Photography.

While covering only 0.8 acres, this downtown city park serves a crucial recreation and performance role. To create the park, the designers had to overcome serious obstacles. Much of it sits atop an underground vault space which housed locker rooms for a former ice-skating rink, and there are extensive city and private subsurface utility installations. The vault roof slab is only 11 inches below grade. Green-roof technology was used to provide low-growing shrubs, ground covers, and gardens. To meet the city's need for a downtown performance venue, a multipurpose terrace was designed to provide a stage for events and shaded seating for everyday users. A prominent trellis supports light and sound rigging, with necessary electrical service, controls, and storage in renovated vaults below – along with pumps and filters for the central fountain. Trumpet vines will grow up the trellis to provide shade. Park layout, planting, and lighting are designed to maintain sight lines to the performance terrace. Materials such as brick, cast limestone, and exposed metal reflect the traditional architecture of the community, while responding to the Modern design of the embracing building. With 42 percent impervious ground surfaces – vs. 82 percent previously – the design greatly reduces the amount of reflected light and heat, as well as storm water runoff. In what is understood to be first for such a public urban space, the park employs only organic processes – such as living micro-organisms – to maintain the soil and plantings, with no synthetic herbicides or pesticides, and a plaque at a park entrance explains these organic advantages to the public.

Derck & Edson Associates

Culinary Institute of America Anton Plaza
Hyde Park, New York

The project originated in the Institute's need for parking facilities with handicapped accessibility at the doorstep of the campus's most prominent building. What was once a rocky hillside with a 30-foot drop has been transformed into a 32,500-square-foot plaza above a 145-space, two-level parking garage. And the campus now has a viewing platform worthy of its Hudson River setting. The Culinary Institute differs from other educational institutions by introducing a new class and graduating another one every three weeks, with due ceremony. It is also a destination for over 250,000 visitors a year, who tour the school and dine in a variety of student-staffed restaurants on campus. Creating a plaza atop the garage involved a number of challenges: landscaping with 12 inches of soil depth; providing enough hardscape surface for events such as graduations and banquets along with generous greenery; placing vertical elements – the pergola and the zero-grade fountain – on top of the garage; providing space and support for a 70' x 70' tent for certain events; accommodating vehicles required for maintenance and erecting tents; allowing for the exceptional winds and cold temperatures of the Hudson River site. The plaza is required by the state to act as a storm water filter and does so with engineered soil that does not impose an excessive load on the parking structure beneath. Planting urns, which are changed

Far left: Students posing on plaza.

Left: Evening event on plaza.

Bottom: Plaza overlooking the Hudson.

Photography: Nathan Cox Photography, The Culinary Institute of America, Jessica Paddock.

Derck & Edson Associates

seasonally, provide accents. The axial formality of the plaza design was determined not only to respect the Classical Revival architecture of Roth Hall, which rises above it, but also by the formality of activities at the Institute. Proper kitchen uniforms with toques are required for classes, and formal wear –including long gowns—is common at special events. The zero-grade fountain provides a focal accent for everyday use, but is shut down to leave a flat surface for events. So appropriate is the plaza to its setting that many say it looks as if it has always been there.

Top: Fountain after dark.
Left: Plaza with Roth Hall backdrop.
Photography: Nathan Cox Photography.

Duany Plater-Zyberk & Company
Architects and Town Planners

Miami Main Office
1023 SW 25th Avenue
Miami, Florida 33135
305.644.1023
305.644.1021 (Fax)
www.dpz.com

Washington, DC Regional Office
320 Firehouse Lane
Gaithersburg, Maryland 20878
301.948.6223
301.670.9337 (Fax)

Charlotte, NC Regional Office
119 Huntley Place
Charlotte, North Carolina 28207
704.948.8141
704.948.8144 (Fax)

Duany Plater-Zyberk & Company

Seaside
Walton County, Florida

Founded in the early 1980s, Seaside has long been known as a pioneering demonstration of the compact, mixed-use, walkable community, but few have recognized that it is an environmental landmark, as well. To begin with, all development on the 80-acre site was set back behind the natural dunes, before legislation made this mandatory. The street layout and the vernacular architecture promote natural ventilation, and most storm water is directed to the central square, which acts as retention pond at times of heavy rainfall. Disturbance of natural topography and soil percolation was minimized by supporting houses on pier foundations. Brick-paved roads and footpaths, and gravel swales for on-street parking, are permeable to rainfall. Also, except for civic gathering spaces, all landscaping is strictly limited to native vegetation. With LEED standards decades away, Seaside's early planning demonstrates how age-old strategies can address today's most pressing environmental concerns.

Top: Brick and gravel walkway leading toward one of the pavilions that mark stairways to beach.

Above middle: A typical street in the west end of town.

Above: The first two cottages with screened porches, undercrofts and other vernacular details to maximize cross-ventilation and protection from the elements.

Above right: View over a typical mid-block pedestrain walkway.

Above left: Town master plan.

Bottom left: Aerial view of the town square from the Gulf and its axial relationship to the three principal greenways.

Photography: Courtesy of DPZ, except aerial by Alex Maclean/Landslides.

Duany Plater-Zyberk & Company

Alys Beach,
Walton County, Florida

Alys Beach was conceived and planned as an explicitly sustainable resort town. Sustainability is promoted by placing the most intensive development in compact, mixed-use, walkable neighborhoods, while treading lightly on the more environmentally sensitive portions of the 158-acre site. Pine forest, sand dunes, and approximately 20 acres of wetlands are preserved and integrated into the community as a coherent network of wildlife habitat and open space. At the building scale, the white walls and roofs of Alys Beach's structures give the town an elegant visual simplicity while reducing the heat island effect and improving energy efficiency. Streets are laid out to channel cooling ocean breezes through the community, and most homes use geothermal heating and cooling systems. An energy code requires energy-efficient windows and insulation, and up to 75 percent of generated waste is reused or recycled. A full-time on-site environmental program manager monitors the community's environmental conditions, ensuring the use of green building materials and proper management of construction waste.

Top left: Typical Alys courtyard with fountain.

Top middle: Characteristic architecture blending lessons and influences from Bermuda, the Caribbean and Guatemala.

Top right: Town master plan.

Right: View from a greenway of the first set of houses built along a pedestrian thoroughfare that leads to the beach.

Below: View of a narrow courtyard with a side loggia and breezeway.

Photography: Courtesy of DPZ and The Town of Alys Beach.

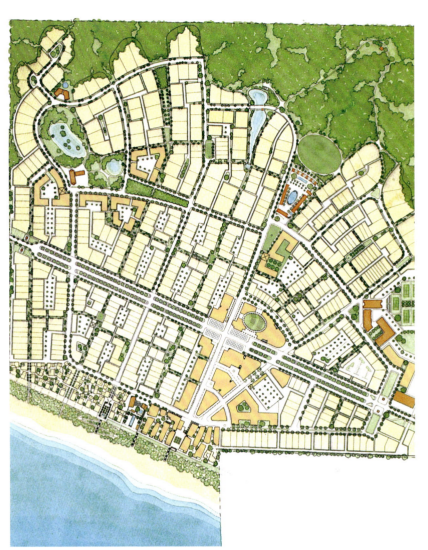

Duany Plater-Zyberk & Company

Schooner Bay
Great Abaco, Bahamas

Schooner Bay is a seaport community modeled after historic Bahamian harbor settlements such as Hopetown and Dunmore Town, and intended as a precedent for compact, mixed-use development on Abaco Island. The master plan for the 220-acre site proposes two villages clustered around a harbor and marina, and delineated by a central greenway. This greenway is in part determined by preservation easements, and maintains the migratory path and foraging environment of the Bahamian parrot. Dunes and vegetation along the beach, as well as forested woodlands, will also be maintained. Developed areas will feature native species requiring minimal maintenance and irrigation. Storm water will be collected and distributed on-site through natural swales, and water retention techniques will be implemented throughout. Material excavated from the harbor will be used for raising and enhancing the topography at the lowest portions of the site to help prevent flooding during hurricanes. The community's architectural code will specify orientation, shading devices, natural ventilation, and construction with green materials, while offering incentives for LEED certification.

Top: Master plan.
Middle: Panorama of the community, with the harbor at the upper central portion.
Bottom: Detail of waterfront and marina harbor.
Renderings: David Carrico.

Duany Plater-Zyberk & Company

Sky
Calhoun County, Florida

Right: Town master plan.

Below left: Aerial view of main village.

Below right: Typical courtyard compound at the equestrian center.

Bottom: Intersection of two rural roads with a cluster of hacienda estates around a green.

Renderings: James Wassell.

By featuring innovative green design measures at both the community-wide and architectural scale, this new town offers a model of growth for a predominantly rural county. Sky clusters its development in compact, walkable villages and hamlets within a natural and agricultural environment. Three villages and two hamlets along the perimeter of the 571-acre site include residential, retail, and office components, while 259 acres remain as open space, of which 150 are dedicated to community agriculture, including gardens, orchards, and vineyards. Sky's buildings combine vernacular building traditions with high performance, environmentally efficient technologies. Environmental components include green roofing, off-the-grid technology, and passive and active solar technology. Cisterns will collect rainwater, and drinking water will be drawn from a local aquifer via on-site wells. The State of Florida's Environmental Protection Agency awarded the town a $1.8-million grant to test energy-efficient technologies and innovative land use practices, and the development team is collaborating with scientists and researchers from two Florida universities.

Duany Plater-Zyberk & Company

East Fraserlands
British Columbia, Canada

Located about 10 kilometers from downtown Vancouver, this 130-acre brownfield parcel was the location of a sawmill that severely damaged its site. Following local policy goals, the plan proposes the redevelopment of the tract as a high density, mixed-use community that promotes sustainability and helps to restore the natural environment. The plan is comprised of a town center neighborhood and two sub-neighborhoods defined by greenways that run the length of the site and connect to a regional green network, thus restoring the site's hydrology. Small foreshore islands enhance fish, wildlife, and bird habitats. A pedestrian promenade lines the site's entire river edge, and the town will have multiple vehicular, bike and pedestrian connections to the greater community. A future transit station is planned along the existing railway traversing the site. A sequence of smaller green spaces, community produce gardens, and pedestrian mews also contribute to the public realm. The plan capitalizes on the site's southern exposure by orienting the buildings and blocks to capture sun for heating and breezes for cooling. A target of LEED Silver rating for buildings will further reduce the town's environmental impact and yield long-term savings for residents and businesses.

Top: Town master plan.
Middle: Aerial view of site within context (drawing courtesy of James Cheng Architects & Associates).
Bottom left: Medium density buildings framing formal greenway (rendering by Pat Pinnell).
Bottom right: View of naturalistic greenway corridor separating neighborhoods (rendering by Laurie Brown).

Duany Plater-Zyberk & Company

Upper Rock District
Rockville, Maryland

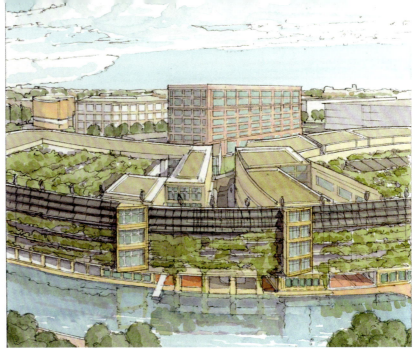

The Upper Rock District transforms a suburban office park into a dense, mixed-use, environmentally advanced, transit-oriented town center. The 20-acre infill site is limited in size but strategically located north of Washington D.C. It is less than one mile from the Shady Grove Metro stop and within walking distance of a proposed light rail station. During a 2002 design charrette, Duany Plater-Zyberk & Company developed two possible plans for the site while exploring different approaches to block structure and spatial definition. The emphasis for both plans is on green architecture and on generating energy onsite via a sound wall replete with solar collectors and a wind harvester. Both plans support a wide variety of uses, including converted loft units, live-work buildings, affordable incubator markets, moderately priced dwelling units, and senior housing. But the urban patterns of the plans are quite different: the first based on a more orthogonal grid with geometrically regular public spaces, and the second based on a more medieval and organically picturesque layout. In both plans, a new LEED Silver eight-story office building is well integrated into the new block structure and is flanked by smaller structures forming pedestrian-scale civic spaces.

Above: Sound wall along highway frontage, with solar collectors and wind harvester.

Above left: Plan based on orthogonal principles.

Left: Plan based on medieval, organic precedents.

Bottom left: Aerial view with highway interchange in foreground.

Duany Plater-Zyberk & Company

Marineland
Flagler County, Florida

Originally the site of a 1930s roadside aquarium attraction, Marineland became a 160-acre incorporated town in 1938. In 1988 DPZ assisted in the creation of a Comprehensive Master Plan, and a charrette in 2002 expanded the vision to establish a model urban and environmental community. Recognizing the site's natural assets, the open space network includes an assortment of storm water retention parks and ponds, as well as a series of communal gardens. A half-mile of oceanfront and inland waterway coastline would remain public, along with 90 acres of live oak hammock designated as a preserve. The built infrastructure would be kept very light, with the vehicular network reduced to the fewest thoroughfares possible, which can be done in resort communities. The majority of the east-west connectors would be pedestrian passages that allow easy access to the beach or the woodland preserve along the Intra-coastal Waterway, while also inviting cooling breezes. Parking in single-family areas would be centralized in parking courts, which feature structures known as "carburetors" containing state-of-the art energy-generating technology designed to serve the entire community, such as wind mills, solar panels, and rain-collecting cisterns. The use of recycled materials and off-the-shelf components was recommended to promote affordable housing. Finally, Marineland's urban and architectural code was drawn up to ensure sustainability in perpetuity.

Top: Town master plan.
Upper middle: Typical pedestrian walkway.
Lower middle: Aerial of Greenway spine with parking 'carburetor'.
Bottom: Larger residences near parkway along intracoastal waterway.
Renderings: James Wassell.

Ehrenkrantz Eckstut & Kuhn Architects

23 East 4th Street
New York, NY 10003
212.353.0400
212.228.3928 (Fax)
www.eekarchitects.com

Washington, DC
Los Angeles
Shanghai

Ehrenkrantz Eckstut & Kuhn Architects

Battery Park City
New York, New York

Right: Master plan of Battery Park City and image of Battery Park City today.

Below: Portion of riverfront Esplanade, showing details reflecting New York park traditions.

Ehrenkrantz Eckstut and Kuhn Architects' (EE&K Architects) master plan for Battery Park City created an entirely new urban neighborhood on 92 acres of waterfront landfill. The plan established a grid of streets at historical Manhattan scale, with a series of parks and other focal points. A stringent set of design guidelines provided architects of individual parcels with the means to ensure sympathetic relationships and capitalize on river views. The development has won numerous prestigious awards, both as a plan and as it has moved incrementally toward completion. One essential feature of the plan is the 1.2-mile Esplanade, which maximizes public enjoyment of the riverfront. It provides cloistered settings as well as a generous pedestrian thoroughfare and is studded with public art by world-famous artists. In one area, the Esplanade leads into South Cove Park, which introduces boardwalks, jetties, a viewing platform, a bridge to a floating island, and a distinct palette of plant materials – all to bring the public into closer contact with the water and harbor views. Among the numerous commercial, residential, and mixed-use buildings that contribute to Battery Park City's urban quality, EE&K Architects has designed the Liberty View Apartments, a 28-story, 300,000-square-foot structure in the South Cove vicinity, designed to exemplify the guidelines established by the firm in its master plan, with apartments oriented to views of the river and the Statue of Liberty. EE&K Architects is currently designing two residential buildings anchored by a community recreation center. These buildings will be built on the last available site of Battery Park City.

Above: Liberty View Apartments at center, with elements of South Cove Park, designed with artist Mary Miss, in foreground.

Left: Aerial View of Rector Place in Battery Park City.

Far left: Gardened square within Rector Place.

Photography: Stan Ries.

Ehrenkrantz Eckstut & Kuhn Architects

Arverne-by-the-Sea
Arverne, New York

Top left: Completed Phase 1A shorefront housing.

Top right: Beachfront of planned Tides Neighborhood.

Above: Site plan of entire 100-acre Arverne-by-the-Sea.

Left: Portion of development centered on Ocean Way, main transit-to-beach thoroughfare.

Photography: Taylor Photo.

The firm's plan for this 100-acre site capitalizes on its unique location between a subway line and the Atlantic Ocean in New York City's borough of Queens. Despite its advantageous setting, the area has resisted redevelopment for over 30 years. This transit-oriented development is planned to include 2,300 units of housing, plus retail and a host of public facilities set within walkable neighborhoods, each with a range of housing types, views of the ocean, and access to the beach. Ocean Way is designed as the main street, linking transit and oceanfront, with a mixed-use environment for residents and visitors. Street-level retail will include both beach-oriented retail such as surf shops, ice cream shops, and restaurants and year-round establishments. Phase 1A includes four housing types designed by EE&K Architects, including two- and three-story semi-detached units. Taking cues from Atlantic coastal vernacular, the houses feature covered porches, screened balconies, bay windows, and rooftop terraces. Landscaped "mews" streets knit the 64 new homes together around the semi-private Dunes Garden, which is planted with native seaside species. The Tides Neighborhood is planned to include 884 dwelling units and 134,000 square feet of retail space. Oriented to Ocean Way and views of the Beachfront Plaza, the plan includes both apartment buildings up to 11 stories high and townhouses along an intimately scaled lane.

Ehrenkrantz Eckstut & Kuhn Architects

New River at Las Olas
Fort Lauderdale, Florida

This mixed-use development along the banks of the New River will transform the Fort Lauderdale skyline and provide a new, distinctive destination. The complex provides a crucial link between the downtown, which has experienced significant growth recently, and both the waterfront and the historic district. Within its 1,822,000 square feet of construction will be 253 condominiums, a hotel and conference center, office space, a theater and an art-oriented cinema, and street-front retail. The totals of residential and commercial space are governed by the city's downtown master plan and agreements with the city that include negotiated requirements for structured parking inside the complex.

Of the development's 3.85 acres, 1.11 acres will be public space, including a new public plaza and promenade along the river and a pedestrian street leading to the river through the center of the project. Of the development's three buildings, the two along the river are curved to expand the public open space. The buildings will feature green roofs, with planting and activity areas, and sun-shading techniques will mitigate heat gain through glazing.

Top left: The complex seen in context of city and river.

Middle left: Aerial view showing relationships of project's three buildings along pedestrian street.

Bottom: View from river, showing two curved buildings defining public plaza.

Computer renderings: Blue Presentation.

Ehrenkrantz Eckstut & Kuhn Architects

Gateway Center
Los Angeles, California

Los Angeles, known for its freeways, is not where one would expect to find the country's largest intermodal transportation facility, also serving as the centerpiece of a pedestrian-focused revitalization plan. But EE&K Architects' Gateway Center and master plan for the 75-acre Alameda District provide both. The goal was to return this neglected location adjoining the city's historic Spanish Revival railroad station to its origin as the grand portal to Los Angeles. Gateway Center is organized around Metro Plaza, as the connecting point between

transportation modes as well as a generously landscaped pedestrian space. Laid out on two levels, the plaza serves as an outdoor waiting room, with a ramp, elevators, and stairs making essential vertical connections. Inspired by arroyos – the dry riverbeds of the region – Metro Plaza offers shaded cloisters, verandas, and cool fountains, along with public art. And it provides access to a four-level, 3,000-space underground park-and-ride facility. The Gateway Intermodal Transit Center, entered through the stately East Portal, offers access under one roof to every means of ground transportation, including cars, buses, subways, commuter rail, and Amtrak. The Portal's stone archway on Metro Plaza presents a welcoming façade for commuters and recalls the arches of Union Station, to which it is linked by an underground concourse. Inside, a dome of steel and etched glass rises to a height of 90 feet above a lively scene of newsstands, cafés, sweeping murals, a 7,000-gallon aquarium, and a setting for art exhibits and live performances.

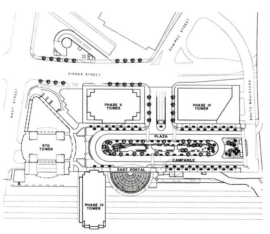

Left: East Portal seen from Plaza.

Top middle: Aerial view of complex, with future tower sites flanking plaza and Metropolitan Transit Authority headquarters at far end.

Top right: Interior of East Portal.
Above: Site master plan.
Above right: Metro Plaza walkway, showing stairs, elevator, and canopy over seating area.

Photography: Walter Smalling.

111

Ehrenkrantz Eckstut & Kuhn Architects

Hollywood & Highland
Los Angeles, California

EE&K Architects incorporated both the historical and the legendary elements of Hollywood Boulevard's heyday in this 1.3-million-square-foot mixed-use development, which includes retail, entertainment, and residential components. A principal goal was to revive the original pedestrian-friendly streetscape, with restrained facades that respond to rhythms, massing, signage, and ornament of early 20th-Century neighbors. Inside the project's precincts, visitors experience a series of forecourts and outdoor stages, linked by distinctive promenades. To maintain the pedestrian character of the boulevard, the vehicular entrance is located on the side street, through a courtyard landscaped as an orange grove. Linked to this courtyard, Actor's Alley is a paseo with a back-of-the-house atmosphere. Flowing up from the boulevard, a monumental stair leads to the Babylon Court, the central public space designed as a recreation of the epic set of D. W. Griffith's "Intolerance." Here a dramatic arch frames a view of the landmark Hollywood sign on the hillside above. Further along the boulevard, a portal leads to Orchid Walk, a processional arcade with a "red carpet" of terrazzo leading to a central rotunda and, beyond, to the state-of-the-art Kodak Theater, the new home of the Academy Awards.

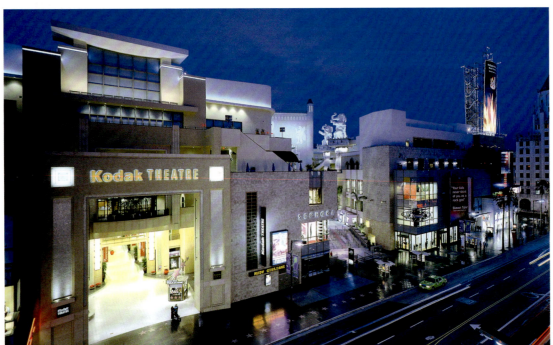

Top left: Corner at Hollywood and Highland, with towering sign.

Middle left: View from Highland, looking toward Hollywood Boulevard, with entry passage at right.

Middle right: Rendering of entire development.

Left: Development's frontage along Hollywood Boulevard.

Photography: Courtesy of Ehrenkrantz Eckstut and Kuhn Architects.

Hughes, Good, O'Leary & Ryan, Inc.

6 Executive Park Drive
Suite 300
Atlanta, GA 30329
404.248.1960
404.248.1092 (Fax)
hgor@hgor.com

Hughes, Good, O'Leary & Ryan, Inc.

Magnolia
Charleston, South Carolina

This 125-acre project is the largest-ever planned development in Charleston and the first major application of the city's pioneering "gathering place district" concept. With an internationally admired pedestrian-friendly historic core, Charleston is encouraging New Urbanist developments as alternatives to sprawl in its outer districts. Required under "gathering place" rules to reserve 10 percent of its area for public opens spaces, Magnolia will exceed that with 16.5 acres of parks and open spaces, plus generous tree-shaded streets. Some streets will have canals in their median park strips, which will play a role in storm water management. Buildings must be configured to maintain two-story-high street walls, with limited interruptions. Park areas bordering the marshes along the Ashley River will be developed with gardens, boardwalks and overlooks, along streams and bio-swales designed to treat storm water. A wide variety of residential units will be located in walkable proximity to office and retail facilities.

Top left: View of development massing, looking west toward river from highway.

Top right: Open space plan.

Right: Boulevarded street with canal in median park area.

Opposite: Site plan, showing river to west and I-26 to east.

Images: HGOR, Inc.

Hughes, Good, O'Leary & Ryan, Inc.

Allen Plaza
Atlanta, Georgia

HGOR carried out the planning and urban design for a high-density, mixed-use development on the northern edge of downtown Atlanta and designed its streetscapes and public open spaces. The six-city-block site now consists largely of surface parking. Great importance was placed on pedestrian connections to the adjacent MARTA transit station and to nearby Centennial Olympic Park, as well as provisions for a proposed trolley route through the site. A key feature of the design is a rooftop terrace accessible from both the proposed W Hotel and the neighboring 55 Allen Plaza (Ernst & Young Building). Integrated planters will offer visitors seating either on benches or on the turf. A series of tensile cable and mesh structures on the terrace will be planted with vines to provide vertical elements without incurring the costs of deeper soil profiles or larger plant materials in this green roof situation. Illuminated at night, these green cones will provide identifying focal points when seen from adjacent buildings and nearby highways. The wall below the terrace will be rendered green by jasmine vines on trellises. The development's streets incorporate the city's streetscape standards, including 10-foot clear zones for sidewalks, a 5-foot zone of special paving for street furniture, lighting, and planting, and accessible curb ramps at all pedestrian crossings.

Top: Proposed W Hotel with roof terrace to left shared with 55 Allen Plaza office building.

Right: Master plan, showing building massing, pedestrian and trolley routes, and green roofs.

Opposite top: View of development from north.

Opposite bottom left: Roof terrace featuring stainless steel tensile structures supporting vines, with vine-covered wall rising from entrance plaza shared by hotel and offices.

Opposite bottom right: Plan view of roof terrace and entrance plaza.

Images: W Hotel rendering by Yen-Ming Lee of Pickard Chilton Architects for Barry Real Estate; other images by HGOR, Inc.

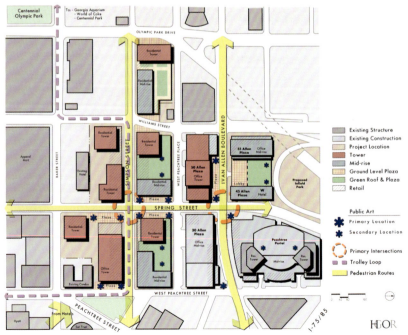

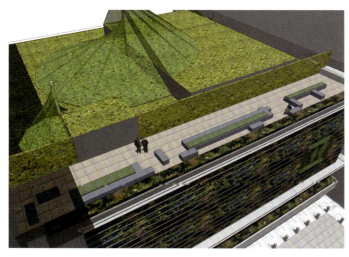

Hughes, Good, O'Leary & Ryan, Inc.

Columbus Riverfront Office Building
Columbus, Georgia

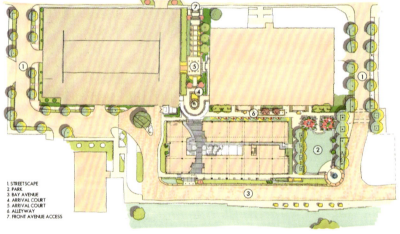

HGOR planned the three-acre development to optimize the relationships among a private office building, a shared public/private parking garage, and a public park. It was essential for the project to capitalize on views of the Chattahoochee River and link seamlessly to the 10-mile river walk, as well as the adjacent downtown retail district. To better accommodate a public street that formerly divided the site, HGOR developed a proposal whereby the private client and the city traded land, improving the relationship between parking and building, and creating a public edge and roadway between the building site and the river. The design of the 100,000-square-foot office building draws on the brick-walled architecture of the area's historic warehouses. Among other advantages, the project offers designated locations for outdoor art —public art at one end and corporate art at the other – which will reinforce the theme of public/private collaboration.

Top left: Park at southwest corner of development.

Middle left: Trellised shelter in park.

Bottom left: Landscaped pedestrian alleyway between new office building and older structure.

Top right: View of Chattahoochee River, with Riverfront Office Building on left bank.

Above: Site plan, with new office building to west, facing river, and parking garage to northeast, adjoining downtown.

Opposite: Office building along new public street, with overlooks to river walk below.

Images: HGOR (top right and above), Bob Hughes (opposite), Ralph Daniel (all others).

Hughes, Good, O'Leary & Ryan, Inc.

Terminus
Atlanta, Georgia

Located at one of Atlanta's key intersections, Peachtree Road and Piedmont Avenue, Terminus will be a mixed-use development incorporating an existing Hyatt Hotel, one million square feet of office space, 200,000 square feet of retail, and 1,400 residential units. HGOR led the design process with a two-day charrette involving architects, engineers, and client to establish the building forms, and drew up a master plan for the entire 27-acre property. The 14-acre Phase I is featured here. It includes a pedestrian corridor called Café Street, which runs from Piedmont Avenue to the roundabout in front of the Hyatt. Lined with high-end retail and restaurants, Café Street will give the Buckhead area a public gathering place for days and evenings, protected from weather by a tall glass canopy. The project will create an environment less dependent on the automobile and establish a benchmark for urbanity in this part of Atlanta.

Top left: Concept rendering of completed project, as seen from Piedmont Avenue.

Above: Rendering of office tower at corner of Peachtree Road and Piedmont Avenue, with canopied Café Street at left.

Left: Plan of 14-acre Phase I, showing pedestrian Café Street crossing site.

Images: Renderings above by Duda/Daine Architects, LLP. Plan by HGOR, Inc.

James, Harwick+Partners, Inc.
Planning • Architecture • Urban Design

8340 Meadow Road
Suite 248
Dallas, TX 75231
214.363.5687
214.363.9563 (Fax)
www.jhparch.com

James, Harwick+Partners, Inc.
Whole Community Design™

Cityville Greenville
Dallas, Texas

Above: Aerial view.

Right: Liner units with leasing office in foreground, existing mature trees along street.

Far right: Avenue frontage, with loft residential units above retail.

Opposite bottom left: interior courtyard.

Opposite bottom right: Site plan, showing townhomes at left, apartments, garage, and retail at right.

Photography: Steve Hinds; Rion Rizzo.

Located in the trendy Lower Greenville area of Dallas, less than three miles from downtown, this urban infill, mixed-use development comprises 128 residential units, 15,000 square feet of retail, and 12 townhomes on 3.65 acres. Compact and pedestrian friendly, the project is woven into the existing fabric, reinforcing the distinctive characteristics of Lower Greenville. Its planning involved meetings over a period of six months with neighborhood groups, dealing with such issues as parking, scale and massing, vehicular access through adjoining residential areas, and quality of materials. The development was then conceived as four sub-areas: a high-density retail zone (1), with loft units above, fronting on Greenville Avenue; a central parking garage (2); a transitional area of three-story multifamily structures (3); a row of townhomes (4) making a transition to the adjacent single-family neighborhood. The retail section reflects the avenue's mercantile tradition; its upper-floor housing is set back from the storefronts. The garage, providing shared retail and residential parking, is screened by buildings and landscaping. The apartment and townhome buildings have warm-colored stucco, sloping roofs, and details such as entry stoops and bay windows that relate to the predominantly bungalow-style houses nearby. Residential amenities include two inner courtyards, which contain a recreational pool and quiet lounge areas. A public courtyard behind the retail row features a decorative fountain and café tables.

James, Harwick+Partners, Inc.
Whole Community Design™

The Commons at Atlantic Station
Atlanta, Georgia

Top left: Park District seen across Grand Ellipse Park.
Top right: Park District streetscape with leasing office.
Above: Portion of Park District.
Left: Aerial view.
Photography: Rion Rizzo.

The Commons is the residential heart of Atlantic Station, a national model of sustainable development that will include 12 million square feet of retail, office, residential, and hotel space and 11 acres of public park. The 138-acre in-town site, formerly occupied by a steel Mill, required $10 million worth of remediation, with the removal of over 12,000 truckloads of contaminated soil. To improve Atlanta's natural environment, the master developer spent $25 million on a state-of-the-art sewage system, the only one in the city that separates sanitary waste from storm water. For the entire development, strategies were established to minimize pollution and ensure sustainable building practices. JH+P has designed two residential sections flanking the five-acre Grand Ellipse Park, Atlantic Station's major recreation and public event space. Landscaping around the park's elliptical central pond includes a 60-foot smokestack and other relics of the steel mill. The firm's two residential areas are the Park District, comprising 231 rental apartments – with 20 percent affordable units – and the Arts District, including 347 condominiums. Both areas are laid out for convenient pedestrian circulation, and no parking areas are visible from the streets. The Park District features a variety of architectural forms, clad in earth-toned materials compatible with the park. The Arts District, designed primarily for young professionals, has a more urban look, with geometrical forms and bold colors inspired by Piet Mondrian's grid paintings.

James, Harwick+Partners, Inc.
Whole Community Design™

Cityville Southwestern Medical District
Dallas, Texas

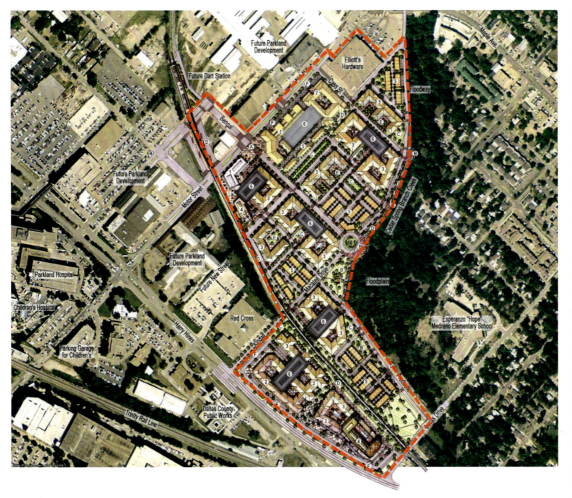

For an area near downtown Dallas, adjacent to the Parkland Hospital campus, this project proposes transit-oriented development (TOD) for a future rail station in the DART rapid transit system. Planning took place on three levels: guidelines for a tax-increment-financing (TIF) district of over 30 acres, a master plan for a planned district of over 15 acres, and design of a 5.7-acre phase one of the development. The first phase will comprise 265 housing units, 15 townhouses, and 40,000 square feet of street-level retail. A retail plaza across the street from the DART station will serve as the entrance to the future neighborhood. Occupying the site of former industrial and warehouse buildings, the greyfield development will initiate a sense of identity and mixed-use community around a transit node.

Top: Study plan of entire tax-increment-financing district.

Left: Rendering of district, with transit station in foreground.

125

James, Harwick+Partners, Inc.
Whole Community Design™

West Highlands
Atlanta, Georgia

This 152-acre project is part of a 500-acre multiphase West Highlands development, planned on new urbanist principles, that replaces a 1,072-unit public housing complex. The master plan calls for both rental and ownership housing, commercial development, a YMCA, a day care center, a public library, a new school, parks, ball fields, nature trails, and a public golf course. JH+P's portions include: the 152-unit Columbia Crest on 3.65 acres at the heart of West Highlands; the 124-unit Columbia Estates on an adjacent 7.14-acre site; and the Columbia Heritage Senior Residences, with 132 units for seniors on a 3.49-acre parcel. Columbia Crest's multifamily buildings, wrapping around and concealing parking structures, contain 1-to-3-bedroom units – both affordable and market-rate. Columbia Estates includes both flats and townhouses, with on-street parallel parking for visitors and resident parking in the rear. It consists largely of affordable units, designed to be indistinguishable from market-rate ones. Resident amenities include swimming pools, fitness centers, and computer rooms. The Columbia Heritage Senior Residences include 60 percent affordable units and are designed to blend seamlessly with other neighborhood residences. The building program included several social spaces and corridors relieved by seating areas with natural light. Architectural design throughout reflects the Craftsman and Bungalow traditions of the area's homes.

Top left: Columbia Estates streetscape.

Top right: Columbia Heritage Senior Residences.

Above left: Aerial view of Columbia Estates.

Above right: Columbia Crest and Columbia Heritage viewed from park.

Photography: Rion Rizzo.

James, Harwick+Partners, Inc.
Whole Community Design™

Museum Place
Fort Worth, Texas

The 10-acre mixed-use development is located on underutilized blocks across from Fort Worth's cultural district, that includes such world-class institutions as the Kimbell Art Museum, designed by Louis Kahn, the Amon Carter Museum designed by Philip Johnson and the Modern Art Museum, by Tadao Ando. To complement these cultural amenities, careful analysis led to a master plan for a revitalized urban neighborhood, including retail, offices, cafés and hotels, along with varied high-density housing types, all sharing several parking structures. Vibrant 24/7 activity is anticipated. The plan also reestablishes the street grid of the neighborhood, providing connectivity and a retail corridor.

Top: Aerial view of proposed development.

Left: Ground-level view of proposed streetscape.

James, Harwick+Partners, Inc.
Whole Community Design™

Cityville Fitzhugh
Dallas, Texas

Left: Principal corner with leasing office.
Above: Aerial view.
Below left: Fireplace in main courtyard.
Photography: Steve Hinds.

The project took a 3.57-acre abandoned office site between downtown and uptown Dallas and transformed it into a mixed-use community with 222 residential units for working professionals of all ages. From master planning through development, JH+P worked closely with the city's planning department to develop a Planned District. Major goals were: emphasis on the street, creating urban edges; preservation of live oak street trees; creating a gateway to the surrounding community; increasing density as appropriate to the surrounding neighborhood; creating pedestrian-friendly streetscapes; uses and building design appropriate to each street; ensuring security through an "eyes-on-the-street" approach to windows, balconies, patios, and driveways. Among the project's attractions are three courtyards, the largest containing a lap pool and generous seating around a stone fireplace, the two others with water features and shaded seating areas. The development has spurred transformation efforts in the surrounding neighborhood.

Joseph Wong Design Associates (JWDA)

2359 4th Avenue
San Diego, CA 92101
619.233.6777
619.237.0541 (Fax)
jwda@jwdainc.com
www.jwdainc.com

Joseph Wong Design Associates (JWDA)

16th/Market Affordable Residential Project
San Diego, California

Left: The 12-story complex seen from 16th and Market.
Middle left and right: Complex model.
Bottom: Open stairs connecting lower apartment floors with second-level courtyard.

Of over 90 residential projects in downtown San Diego now in planning or construction stages, only four focus on affordable housing, and only this one provides for families earning 30 to 65 percent of the area median income. Over 70 percent of the units have two or three bedrooms. The project embodies several family-friendly strategies. Larger apartments are clustered on floors 2 through 5 of the 12-story structure, which are linked by open-air stairways to a 6,900-square-foot courtyard at second floor level. Providing a secure play area for children, it also has a variety of seating areas and lush landscaping, encouraging community engagement and supporting parental supervision. Other resident amenities include a multi-use community room adjacent to the courtyard, laundry facilities, and a rooftop community room opening to a 2,250-square-foot deck. A 6-foot windscreen around the deck assures comfort and safety, while affording views. Each unit has a private balcony with safety features appropriate for children. The first floor of the 166,000-square-foot structure includes 4,700 square feet of retail spaces and residential common areas, including classrooms and facilities for resident support services. There is a two-level subterranean garage.
Scheduled for completion in 2008, the project offers hope that inclusionary housing is an achievable goal through new hybrid relationships between business and aid organizations.

Joseph Wong Design Associates (JWDA)

North County Regional Education Center
San Marcos, California

As the education hub in northern San Diego County for the county's Office of Education and the San Marcos Unified School District, this 71,000-square-foot, $21-million building serves a variety of needs. Along with offices, conference rooms, and a cyber café, it houses regional intervention programs, tutoring, juvenile court school, and a beginning teachers support and assessment program. The facility also helps introduce students and teachers to the latest in classroom technology. Sited among office parks and civic buildings, the structure needed a prominent feature to set it apart. Its 40-foot-high entrance volume, clad in Indian sandstone, gives it a distinctive identity and is well related to its stucco-clad volumes. Energy strategies include double-paned low-e windows with different glazing depending on relation to light shelves. Carpet is made of 100 percent recycled fiber, and paints and coverings have low VOC to minimize out-gassing. The building's layout around a courtyard ensures that floor plates are not too large and interiors receive ample daylight.

Right: Entrance volume after dark.
Bottom: Overall view.
Photography: Jim Brady.

Joseph Wong Design Associates (JWDA)

Lane Field
San Diego, California

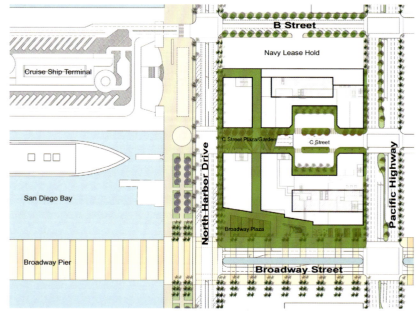

A 5.8-acre property on San Diego Bay will be transformed from parking lots to a generously landscaped retail, restaurant, and entertainment development, with 250 luxury executive suites and a 550-room InterContinental Hotel. Plans recently approved by the Port of San Diego will allow the redevelopment of the former Lane Field site for the accommodation of residents, vacationers, and day visitors. Under the plans, C Street will be converted into a pedestrian promenade that opens toward the harbor and its expansive views. In addition, a 100-foot-wide esplanade along the water – a true waterfront park – will be created by moving Harbor Drive 70 feet to the east.

The plans conform to the North Embarcadero Visionary Plan, which calls for infrastructure improvement and aesthetic enhancements in bayfront redevelopments.

Top left: Development seen from bay, with Broadway Street and Pier at center of view.
Top right: Towers on podiums seen from waterfront esplanade.
Left: Site plan.
Bottom left: Complex seen from city side.

Joseph Wong Design Associates (JWDA)

Sofitel JJ Oriental Hotel
Shanghai, China

This new 400,000-square-foot hotel tower will be linked to an existing hotel through an Oriental garden, with water features and regional landscaping features. While linked to the original structure, the new building will include full-service facilities, including 450 rooms and suites, a restaurant and bar, an executive lounge, and meeting and ballroom spaces. Together, the buildings will form a five-star hotel property in Shanghai's burgeoning Pudong district, overlooking Century Park. The opening is scheduled for 2008, in time for travelers to the Olympics in Beijing. JWDA won the commission through an invited international design competition.

Top left: Aerial rendering of entire hotel complex.

Top right: New tower with low-rise wing and portion of garden.

Above: New hotel building with existing hotel in background.

Left: Site plan, showing garden linking hotel structures.

Joseph Wong Design Associates (JWDA)

Deep Blue Plaza
Hangzhou, China

Right: Night view of complex from river.

Opposite top: Complex seen from river, with curved apartment towers facing south.

Opposite bottom left: Site plan.

Opposite bottom right: Residential lobby.

Photography: Kerun Ip.

This mixed-use project on a choice riverfront site combines two residential towers, a discrete office tower, and a two-story commercial/retail building in a new landmark for the burgeoning city. Visibly related to the river, the iconic curvature of the towers affords all residences south-facing balconies with views of the famous West Lake. Their siting buffers them from the busy adjoining roads and the central business district to the north and allows ample sunlight exposure to neighboring residential developments. The buildings have double-glazed, high-efficiency windows to control solar heat loads. The project's more than 100,000 square feet of public open space, graced with fountains, reflecting pools, and opulent gardens, is linked to the city's pedestrian promenades. Landscaping is given a soft, organic aspect to complement the crisp Modernism of the buildings and offer a

Joseph Wong Design Associates (JWDA)

respite from the intensity of the adjoining roads and business district. The complex includes underground parking for its tenants. The low-rise commercial building north of the residential structures offers space for restaurants, cafés, personal services, and retail shops, which serve both residents and tenants of the office tower.

Top left: Office tower by night.
Top right: Office tower at northeast corner of site.
Right: Residential facades.
Photography: Kerun Ip.

JPRA Architects

31000 Northwestern Highway
Suite 100
Farmington Hills, MI 48334-2585
248.737.0180
248.737.9161 (Fax)
www.jpra.com

JPRA Architects

The Village of Rochester Hills
Rochester Hills, Michigan

An existing enclosed mall with large asphalt parking lots has been transformed into a streetscape with trees, flowers, lamp posts, park benches, and curbside parking. The resulting 375,000-square-foot complex embodies the theory, design, and technology of "Main Street USA," as defined by the Urban Land Institute (ULI). Anchoring the 1,000-foot-long main street is a new 120,500-square-foot department store at one end and 54,000-square-foot food store at the other. A variety of upscale shops and restaurants line the street and plazas between. Individual storefronts of brick and stone, with creative signage and awnings, establish a comfortable pedestrian scale. Landscaped walkways lead from small parking areas behind the retail buildings. The firm's services extended from master planning through building and lighting design.

Top right: Street scene with signage, lighting, and planting.

Middle: View toward new Parisien department store.

Bottom left: Parklet with gazebo.

Bottom right: Main Street with detail of the playground.

Opposite: Main Street with curbside parking.

Photography: Gene Meadows.

JPRA Architects

Somerset Collection South Renovations
Troy, Michigan

Left: Existing sculpture in new setting.

Right: Rotunda with curved, illuminated wall behind new concierge desk.

Below: Fountain with sprays under glass.

Bottom: New glass-enclosed elevator rising from pool.

Opposite: Grand court with central fountain and glass partitions defining seating areas.

Photography: Tom Hurst.

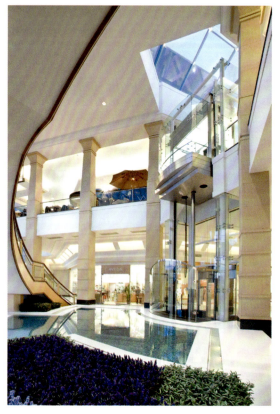

Renovations to the 12-year-old Somerset Collection were carried out to support and enhance its reputation for outstanding programmed events and customer services. The project involved an area of 12,000 square feet, including the entry rotunda and the grand court. In the rotunda, a new concierge desk has been created, backed by a curved wall surfaced with translucent glass and softly back-lit in changing colors provided from LED sources. The café/restaurant that once occupied the grand court has been replaced by a "central park" with seating areas divided by four 10 ft. high glass partitions embossed with wisteria flowers. New floors are of terrazzo, pebble stone, and translucent glass. Existing sculptures have been relocated, and a central fountain with pop-jet sprays under glass slabs provides kinetic interest with subdued sound. New lighting and planters with faux landscaping complete the court's transformation. To increase views of the mall from the rotunda, an existing elevator has been replaced with an all-glass hoistway and cab with mirrored stainless fittings.

JPRA Architects

Palladium
Birmingham, Michigan

Below: Complex in the evening, showing street-level shop windows and central movie marquee.

Opposite top: Complex by day.

Opposite bottom: Quotations and movie-related images on upper walls.

Photography: Laszlo Regos and JPRA staff.

Replacing a former department store at a key point in Birmingham's fashionable shopping area, Palladium is a 140,000-square-foot retail and entertainment development. The complex provides street-level and lower-level retail, plus two upper floors and two mezzanines accommodating a 12-screen cinema with seating for 2,200. Full-height windows in cinema lobbies provide theatergoers with balcony views of the street below. Bold but dignified graphics, pointedly different from signage, add appropriate visual interest to windowless volumes at the

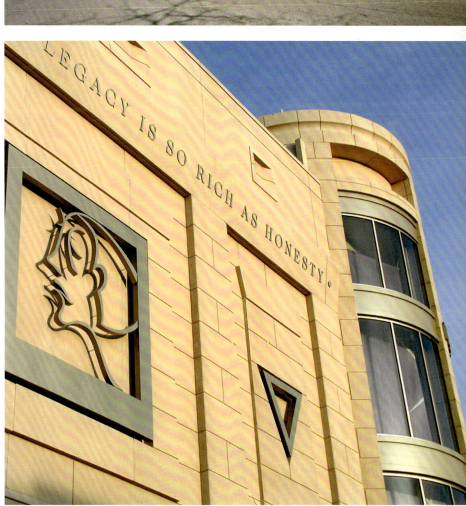

JPRA Architects

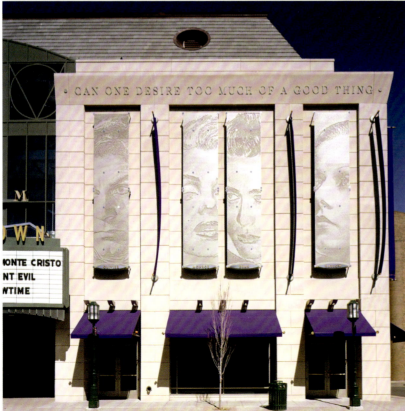

building's two ends. Precast stone façades with custom-detailed windows and awnings reinforce the building's role as a landmark in a prime downtown location. Opened in late 2001, Palladium has re-energized its pedestrian-friendly district.

Above: Opposite end of complex at dusk.

Left: Detail with movie-star images on upper walls: Olivier, Monroe, Dean, and Harlow.

Photography: Laszlo Regos and JPRA staff.

LandDesign

200 South Peyton Street
Alexandria, VA 22314
703.549.7784
703.549.4984 (Fax)
www.LandDesign.com

Washington, DC
Charlotte, NC
Pinehurst, NC
Asheville, NC
Tampa, FL
Nashville, TN
Beijing, PRC

LandDesign

SouthPark Mall
Charlotte, North Carolina

Originally built in the 1960's, this regional mall, located in one of the city's most prominent neighborhoods was repositioned and expanded in 1999. This repositioning has been pivotal in transforming this retail and class A office epicenter into a diverse and upscale destination. LandDesign provided land planning, landscape architecture, civil engineering and played a major role in the place-making vision that created an award-winning mixed-use experience. The core of the project was the phased expansion of the existing shopping mall, adding over 1 million square feet of retail and more than 3,500 parking spaces. Maintaining pedestrian and vehicular circulation was successfully resolved with a real-time access management strategy. Surrounding the traditional mall, an outdoor lifestyle center extends the retail experience. The new West Plaza open-air restaurant court is energized by a centerpiece fountain, while the Village combines trendy shops and eateries with five floors of luxury residential. Southern vernacular elements of brick and wood trellis are combined with a native southeastern landscape. This landscape extends into the 15-acre Symphony Park. This multi-purpose green space with a gently sloping lawn functions as a neighborhood park, as well as being a venue for the summertime Charlotte Symphony series. An aesthetic pond creates a stunning backdrop for events while incorporating significant stormwater management. Along with the park, the creation of a landscaped promenade has given the mall a more outward community orientation—a goal dear to neighbors and local planners.

Top: New mall entrance.

Opposite top: Charlotte Symphony performance in Symphony Park, with main mall structure in background.

Opposite bottom left: West Plaza fountain.

Opposite bottom middle: The West Plaza.

Opposite bottom right: The Village main street valet.

Photography/Illustrations: Rick Alexander and Associates, Inc. (opposite top), Elizabeth Chomas (top, opposite bottom left and middle), LandDesign (bottom right perspective).

147

LandDesign

The Towne of Seahaven
Panama City Beach, Florida

Left: Bird's eye view of development.

Below: West streetscape
Renderings: Genesis (bird's-eye); Dougherty + Chavez Architects (streetscape)

Extending along the Gulf of Mexico and encompassing a freshwater lake, this complete resort and residential experience promises to redefine Panama City Beach. Conceived as a self-contained urban district with a broad variety of services and activities, the "Towne" offers nearly 4,000 living units, a 60,000 square foot conference center and over 100,000 square feet of retail in a walkable environment. A quarter mile of beachfront is linked to a main-street village with numerous shopping, dining, and entertainment opportunities. Other amenities like a lakefront promenade, a grand resort-style pool and over four acres of rooftop gardens, pools, spas and cabanas bring the beach resort feel into the interior of the 51 acre site. A major goal was to create a pedestrian friendly and integrated streetscape along the major beachfront road dividing the site. To reduce vehicular use after the visitor's arrival, circulating trams and public trolley stops are provided. As well as being an amenity for water activities such as water taxis and fishing, the man-made lake provides for storm water treatment. Other significant steps taken to reduce the environmental footprint of the community include preserving wetlands and augmenting existing beach dunes.

Rooftop plantings and shade trees will reduce potential heat-island affects, thus promoting energy conservation. The planting of native species will reduce the need for irrigation, which will be supplied from the lake and existing on-site wells. Asphalt and concrete existing on the site will be recycled as road base and fill material.

LandDesign

Downtown Silver Spring
Silver Spring, Maryland

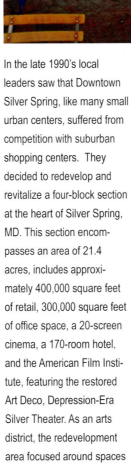

Top left: View of the interactive fountain in the evening.

Top right: Silver Plaza, showing sculptural stair, elevator tower, circular interactive fountain with mosaic tile artwork by Dierdre Saunder of Artbridge.

Middle left: View of restaurants and dining terrace.

Bottom left: Visitors enjoying Silver Plaza and the interactive fountain.

Photography: PFA Silver Spring, LLC (top right, middle left), LandDesign (top left, bottom left).

In the late 1990's local leaders saw that Downtown Silver Spring, like many small urban centers, suffered from competition with suburban shopping centers. They decided to redevelop and revitalize a four-block section at the heart of Silver Spring, MD. This section encompasses an area of 21.4 acres, includes approximately 400,000 square feet of retail, 300,000 square feet of office space, a 20-screen cinema, a 170-room hotel, and the American Film Institute, featuring the restored Art Deco, Depression-Era Silver Theater. As an arts district, the redevelopment area focused around spaces that engaged the senses and encouraged visitors to visualize the area as dynamic, exciting, and inspiring. LandDesign refined the designs of the streetscapes, Gateway Plaza and Silver Plaza. Gateway Plaza replaced an earlier shopping center but preserved its Art Deco façade. A water feature identifying Downtown Silver Spring stands as a landmark and sets the tone for the flowing character of the space. At the heart of Downtown Silver Spring lies Silver Plaza, an events plaza designed for outdoor performances and cultural activities. It is surrounded by outdoor dining terraces and retail shops shaded by light-filtering honey locust trees. Silver Plaza features a dramatic staircase and a dynamic interactive fountain with fiber optic illumination and computer-controlled water jet patterns. Coordinating with artists and designers, tile mosaics were created to enliven the plaza. Downtown Silver Spring has become one of the most successful mixed-use urban districts in the Washington, DC area.

LandDesign

Shenzhen Bay Seafront Urban Design
Shenzhen, China

The subject of this design consultation is a 15-kilometer (9-mile) stretch of bay shore bordering the burgeoning tropical city of Shenzhen. The application of international best practices in urban waterfront development created a sustainable coastal leisure district plan. The project is divided into five distinct zones. Zone A, the leisure residence district, provides areas for culture and sports, while reinforcing the natural beauty of the shoreline. Zone B, the tourism and vacation commercial district, accommodates public gathering, performance, and a water-friendly play area, as well as a mangrove forest ecology garden, a revolving restaurant, a TV tower, and a waterfront amphitheater. Well connected to public transportation, this zone is organized along an axis through a gated central building and adjoining plaza. Zone C, the greenbelt for Shekou Port district, is an ecological corridor linking the D and E zones, designed in a naturalistic style with all tropical plantings. Zone D is a tourist and leisure area, extending the existing Shekoushan Park to the bayfront with a tropical urban landscape. It includes a hotel and a ferry terminal. Zone E includes the Sea World theme park, along with a residential district, a fishing village, a seafood market, and a gourmet dining center. Throughout the five zones, a shoreline that was degraded and eroded is reestablished as ecologically healthy, with minimal but effective use of landfill and over-water constructions.

Right: Zone B, one of five shore line zones, extending from city along central axis and including (left to right) mangrove forest ecology garden, TV tower, and revolving restaurant.

Below right: Tropical plantings and beaches in Zone C.

Bottom right: Offshore bridges and platform in Zone A.

Opposite bottom: Zone A shoreline at twilight.

Illustrations: Shenzhen Dizhijing Landscape Co. Ltd.

LandDesign

Cool Springs Mixed-Use Development
Franklin, Tennessee

On a 284-acre tract currently zoned primarily for office use, Crescent Resources recognized that a single-use development was not what the Franklin, TN market desired. LandDesign's master plan envisioned a mixed-use environment that would enhance pedestrian interconnectivity and create memorable spaces. Two mixed-use centers are proposed: a regional one, close to a planned road interchange, and another that accommodates lifestyle/entertainment uses, including a performing arts center, a hotel, and neighborhood services. Office functions are located in three distinct areas, each close to urban-density housing and convenient to the two mixed-use centers. Lower density housing occupies areas farthest from these centers, but still within walking distance. Public green areas line the waterways that make sweeping curves through the community. Following New Urbanist principles, major streets interconnect effectively with those of neighboring areas. Streets and a trail system are designed to encourage walking. Trolley services will be coordinated with the City of Franklin. Besides reduced dependence on cars, other sustainable features include storm water treatment with bioretention ponds and reduction in parking footprint through shared parking. The total development is planned to include 920,000 square feet of retail, 1.5 million square feet of offices, 902 residential units, 325 hotel rooms, a 40,000-square-foot performing arts center, plus church and civic building sites.

Top: Entertainment & Lifestyle Plaza, office building beyond.
Above right: Urban residential complex overlooking park along stream.
Right: Master plan.
Illustrations: LandDesign.

Lessard Group Inc.

8521 Leesburg Pike
Suite 700
Vienna, VA 22182
703.760.9344
703.760.9328 (Fax)
www.lessardgroup.com

Lessard Group Inc.

Midtown Reston Town Center
Reston, Virginia

Located in the heart of Reston Town Center, which is internationally recognized for its urban planning, this complex includes the district's tallest residential towers. On a site designated for high density along a pedestrian-oriented main street, the project includes two 21-story condominium towers and a single 21-story apartment tower. They rise from a podium that includes a two-story lobby above underground parking levels that accommodate 1,179 cars. Shops and restaurants in this podium reinforce the continuity of the street's retail activity. Resident amenities include a courtyard pool, a private theater, ballrooms, a fitness center, and game rooms. The two towers at either side of the podium are asymmetrically designed, each a mirror image of the other, thus reinforcing the symmetry of the overall composition. Seen from the Dulles toll road, the towers form a prominent landmark. The 295 residential units in each tower introduce open plans, balconies, and sunrooms that take full advantage of spectacular views. Typical floors have eight units, averaging 1500 square feet. There are larger units, including two-story penthouses, on the top floors. The adjoining Alexan at Reston complex includes 698 units in two four-story structures wrapped around a 124,440-square-foot parking garage. One of these buildings contains 2,300 square feet of retail and the other 4,000 square feet of space for the Greater Reston Arts Center.

Top right: Podium at base of tower continuing street's retail and restaurant activity.

Middle: Site of Midtown Reston and Alexan at Reston.

Right: Alexan at Reston residential structure concealing multistory garage.

Opposite: Flanking a multiple-use podium, twin 21-story residential towers, with a third similar high-rise adjacent.

Photographer: May Siu.

Lessard Group Inc.

Clarendon Park
Arlington, Virginia

The 4.13 acres of Clarendon Park are part of a 10-acre redevelopment site once covered with buildings and paved areas of a Sears Roebuck outlet. The project's residential units are physically connected to a mixed-use building and offer convenient pedestrian connections to the adjoining retail center. The plan is composed of townhouses with vehicular access via single-sided alleys, backing up to the mixed-use structure. Residences face community courtyards, one of which offers pedestrian access to the retail center via an attractive walk-through tunnel. Exteriors are designed to complement the predominantly single-family, Craftsman bungalow style houses of the surrounding area. Unit entries and wrap-around porches on corner units reinforce the compatibility with the community. Civil engineers worked closely with the design team to reduce impervious areas and create innovative underground water quality and quantity treatment facilities to minimize impact on downstream drainage systems.

Right: Courtyard landscaping on sloping terrain.

Bottom right: Townhouses with details recalling neighboring Craftsman-style bungalows.

Bottom left: Pedestrian access via courtyard.

Opposite top: Corner townhouses where complex faces surrounding community. Photography: Boris Feldblyum.

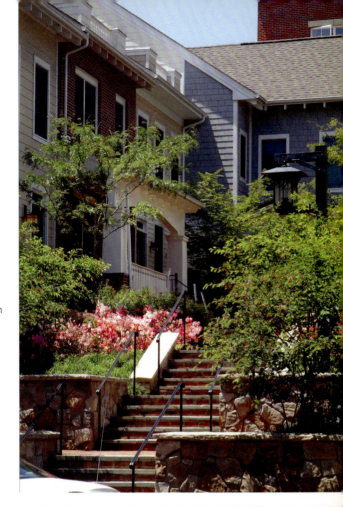

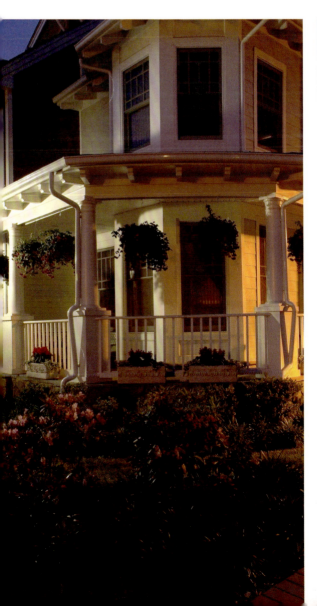

Lessard Group Inc.

Port Imperial
West New York, New Jersey

Occupying about seven acres of Hudson River waterfront, this is part of a 200-acre community planned for a former railroad yard site. Its 432,000 square feet include high-rise and mid-rise apartments. The design was guided by requirements for building heights and massing and for the continuous river walk along the site's frontage, which offers convenient pedestrian access to Manhattan-bound ferries.

Generous amenities and landscaped open spaces give a strong sense of community. All apartments are hard-wired with a comprehensive communications network including telephone, high-speed internet, and home theater systems.

Bottom left: Entrance circle and clubhouse with clocktower.
Below: Port imperial interior.
Bottom: Pool deck with view of Manhattan Skyline.
Photography: Fred Forbes.

Lessard Group Inc.

Chatham Square
Alexandria, Virginia

Located in the historic district of Old Town Alexandria, this innovative mixed-income community occupies the site of a former public housing project. It is a resourceful use of land, replacing 100 public housing units with 100 market-rate, for-sale homes and 52 low-income public housing rental units. The challenge was integrating market rate and public housing and fitting it all into a generally high-income community. The design of the buildings had to be compatible with its historic district and was subject to strict guidelines. The previous two-story structures arranged in a "barracks" pattern were replaced by higher-density three- and four-story townhouses and "two-over-two" apartment buildings designed to look like townhouses. Yet public open space was increased, in a formal, pedestrian-friendly plan with central courtyards. The parking demand is met by garages within the privately-owned homes and underground parking below the two-over-twos, all accessed from rear alleys. The development process was a model of public/private partnership, with proceeds from the sale of public land for the private houses subsidizing the construction of public units – both on and off the site.

Top left: House fronts reflecting variety of periods in Old Town Alexandria.

Top right: Residential units around community square.

Left: Panorama of open space and residential buildings.

Photography: Boris Feldblyum.

Lucien Lagrange Architects

605 North Michigan Avenue
Chicago, IL 60611-3141
312.751.7400
312.751.7460 (Fax)
www.LucienLagrange.com

Lucien Lagrange Architects

X|O, 1712 South Prairie Avenue
Chicago, Illinois

Right: Townhouses along avenue concealing garage podium.

Bottom right: Site plan, showing planted garage roof and neighborhood park at south end.

Opposite: Towers of 33 and 44 stories, with townhouses along South Prairie Avenue.

Among several towers proposed for the resurgent Near South Side, these condominiums are sited along South Prairie Avenue, offering sweeping views of the lake and the city. The design of the complex is conditioned by its location across the street from the world-renowned Glessner House by H. H. Richardson and the landmark district of 19th-Century mansions around it. The neighborhood context has been addressed by placing townhouses along Prairie Avenue, which screen the garage podium behind them, and by providing a small park at the south end of the site, directly across from the Glessner House. The 33-story south tower and its 44-story north counterpart will contain 663,000 square feet of condominiums; the townhouses, 30,000 square feet; parking and lobbies, 205,000 square feet; and lifestyle center, 13,000 square feet. Environmental impact is reduced by the location being in a pedestrian area, near public transportation, by multilevel parking that reduces the building footprint, and by planting covering the garage roof. The project is being designed for LEED certification.

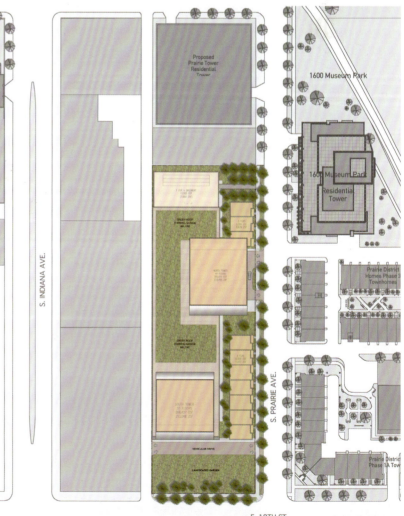

162

Lucien Lagrange Architects

840 North Lake Shore Drive
Chicago, Illinois

For a uniquely favored site facing Lake Michigan to the east and Lake Shore Park to the south, the architects have designed a building that adapts some of the qualities of 19th-Century urban interiors and exteriors in a fundamentally modern structure. The 73 residences in the 368,000-square foot building range from 2,100 to 8,700 square feet, have ceiling heights of nine feet or more high, and enjoy spectacular views from cylindrical bays and recessed balconies. The precast panels of the exterior have been carefully detailed and fabricated to lend distinction to the facades. Zinc-coated copper mansards finish the structure at the skyline. The location encourages pedestrian circulation to public transportation, shopping, recreation, health care and other services. The use of perimeter foundation walls from the building previously on the site saved on resources.

Above: Building's spectacular locale, with lake to right and towers of North Michigan Avenue to left.

Above left: Tower as keystone of its high-rise neighborhood.

Left top to bottom: Precast panels recalling façade details of earlier centuries.

Opposite bottom: Typical whole-floor residence plan.

Photography: © James Steinkamp, Steinkamp/Ballogg Photography, Inc.

Lucien Lagrange Architects

Hard Rock Hotel
Chicago, Illinois

Transforming the Carbide & Carbon Building, a 1920s landmark office building on North Michigan Avenue into a full-service hotel posed architectural challenges. The distinctive terra cotta and brick exterior required thorough restoration, with replacement of nearly 10 percent of the terra cotta. Adapting the office floors for guest rooms called for extensive new plumbing and mechanical systems, supplied through new vertical shafts. A new second fire stair was built and old fire escapes removed. Existing elevators were adapted for guest and hotel staff use, their original doors filled with required fireproofing material. A generous guest reception, lobby, and bar area was created in first-floor space that formerly housed retail tenants. Rehabilitation of the existing elevator lobby included exposing and restoring its original ceiling. The need for public function rooms was met by with a five-story annex, no taller than the building it replaced, along Michigan Avenue. On the inside, continuity was established by aligning floors and using a consistent palette of materials. On the outside, by contrast, the annex appears as a glass-clad work of contemporary design, respecting the integrity of the historic tower.

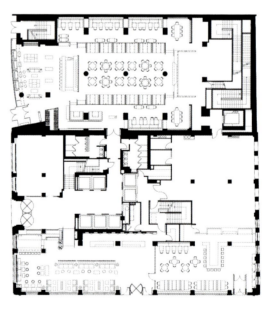

Top left: Lobby in former retail space.

Middle left and right: Addition housing function rooms, with contrasting glass façade.

Bottom left: Restored elevator lobby.

Left: Plans of typical floor and first floor, latter with restaurant in annex at top in plan, lobby and bar at bottom.

Opposite: Thirty-eight story structure, originally Carbon & Carbide Building, rising on North Michigan Avenue.

Photography: © Nathan Kirkman; Scott McDonald © Hedrich Blessing.

Lucien Lagrange Architects

Park Kingsbury
Chicago, Illinois

Planned to be located in the vicinity of two earlier residential towers by the same firm and a riverfront park, this 28-story, 300,000-square-foot condominium will respond to several site constraints. Occupying a triangular site, it is shaped to preserve sight lines from neighboring buildings, following the area's planned development process. The result is a polygonal tower plan, which also maximizes views for its own occupants. The brick masonry relates the building to its industrial context, while its warm grey tone simultaneously relates the building to the contemporary residential towers designed by Lucien Lagrange Architects and other neighboring buildings. Environmental benefits of the project include the site's pedestrian access to parks, restaurants, and other amenities, an on-site park area, and a planted roof over its garage podium. A higher than prevailing ratio of solid wall to window reduces heating and cooling loads.

Top left: Elevation of the polygonal tower rising from garage podium.

Top right: View of area from northeast, this site sharing city block with tower on right.

Left: Site plan highlighting this building, with other recent towers by firm in lighter tint.

MBH

2470 Mariner Square Loop
Alameda, CA 94501
510.865.8663
510.865.1611 (Fax)
www.mbharch.com

1300 Dove Street
Newport Beach, CA 92660
949.757.3240
949.757.3290 (Fax)

MBH

Sonoma Mountain Village
Rohnert Park, California

Sonoma Mountain Village, a 200 acre community proposed on the former Agilent Technologies campus will create a sustainable mixed-use community centered around a vibrant Town Square and will be guided by the principles of deep sustainability, new urbanism and green technology. The homes will range from 800-square-foot condominiums to 3,500-square-foot single-family residences. 90,000 square feet of solar panels have already been installed to power the commercial core. Also an energy center—left behind by Agilent—will be utilized as an efficient heating and cooling system. The community embraces environment-friendly technology that encourages sustainable living. Getting around by car will be an option rather than a necessity since the majority of homes will be located within ¼ mile of the Town Square and a 10-minute walk from the proposed train station. The master planned community includes approximately 1,900 homes, 150,000 square feet of retail, dining, and entertainment choices, and 500,000 square feet of commercial space. Planned amenities include civic buildings, parks, community gardens, an international-size soccer field, a fitness center, as well as miles of trails and bike paths and much more. It is expected to be built over the next 12 years, and construction has already begun to recycle the existing buildings and infrastructure that make up the commercial core.

Top left: Illustrative proposal for retail-office building.
Top right: Characteristic buildings for mixed-use core of community.
Right: Village Square and surrounding development.
Renderings: MBH Architects

MBH

Broadway Arms
Anaheim, California

The first of several mixed-use projects planned for downtown Anaheim, this 172,413-square-foot, five-story structure includes 95 studios, 1-and 2-bedroom units, with retail space on the ground floor and one level of below-grade parking. Units offer such amenities as stainless steel appliances, granite counter tops, and all-in-one washer-dryers. The building includes a generous shared outdoor patio on top of its retail podium. In its exterior expression, the project is articulated as a series of complementary volumes, integrated by a consistent palette of materials. The developer is at work on other buildings in the area that will offer about 600 apartments and condominiums in mixed-use structures. Together, they will generate the foot traffic to support the designation of a stretch of Broadway as The Promenade of Anaheim.

Top right: Building entrance.
Right: Residents' patio above ground-floor retail.
Bottom: View along Broadway, showing rooftop sign.
Photography: RMA Photography

MBH

Broadway Grand
Oakland, California

Sited on a 1.3-acre city block, Broadway Grand is one of several mixed-use residential projects currently under construction that represent an aggressive downtown redevelopment effort by the City of Oakland, California. The building contains 22,000 square feet of ground floor retail surrounding a 3-story parking garage with 4 floors of residential units above, organized around a 4,000-square-foot private courtyard. The 105 flats and 27 townhouses offer spectacular views of the Oakland hills and picturesque San Francisco skyline. Amenities include master suites, stainless steel appliances and granite countertops. Varying façade planes, material and roof heights of the 325,000-square-foot development give passers-by the impression of several smaller buildings that reinforce the scale of the existing streetscape.

Right and below: Views of complex from two different corners.

Renderings: Focus 360.

MBH

Odeon Union Square – 150 Powell Street
San Francisco, California

This mixed-use complex retains the street fronts of a historic four-story corner building and marries its volume to six-story horizontal extensions, which replace a total of four neighboring structures along both streets. The resulting 104,000- square-foot development includes a two-story, 32,000- square-foot retail store, with an additional 12,000 square feet of office and storage space in the basement, along with 29 market-rate residential units. The location on the southeast corner of Powell and O'Farrell streets in the Union Square district offers a variety of nearby public transportation options. The existing un-reinforced brick bearing wall—construction of the historic building from 1906—was replaced with a hybrid structural system

Right: New structures and residential entry on O'Farrell Street.

Bottom left: Historic brick and terra cotta façades with original entry moved to center of Powell Street front.

Bottom middle: O'Farrell Street entrance and canopy.

Bottom right: Fifth-floor lobby.

Photography: Misha Bruk.

MBH

including concrete beams and columns, precast plank floors, and metal framing with polished artisan cement plaster. Because this building is historic, many reviews with city authorities were required for approval to relocate the original entry portal and decorated lobby to the central bay of the main front from its previous off-center location. Elevations of the new wings had to complement the rhythm and cadence of the existing façades, without imitating or overpowering them.

Above: Model residential unit.
Left: Residential units along upper-floor courtyard.

There are four levels of 29 market-rate residential units – ranging in size from 981 to 2,100 square feet. Finishes in units include: hardwood floors, espresso/walnut stained doors and millwork, granite kitchen and bathroom countertops, and high-end stainless steel appliances.

Elizabeth Moule & Stefanos Polyzoides
Architects and Urbanists

180 E California Boulevard
Pasadena, CA 91105
626.844.2400
626.844.2410 (Fax)
www.mparchitects.com

Elizabeth Moule & Stefanos Polyzoides
Architects and Urbanists

The Robert Redford Building for the Natural Resources Defense Council (NRDC)
Santa Monica, California

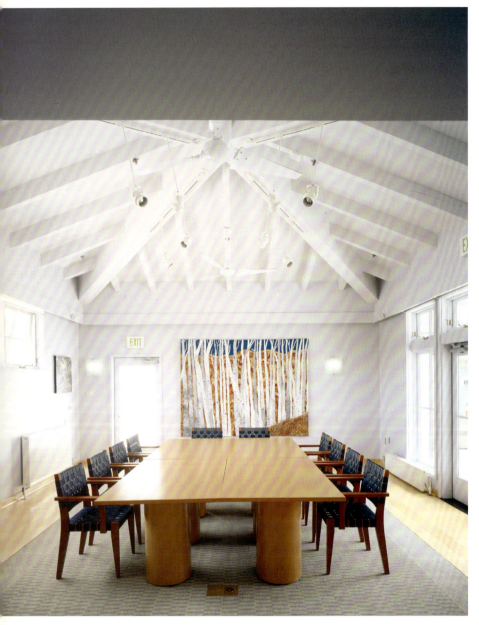

Above: Conference room, with chairs made with recycled seat belts.

Top right: Building front, remodeled from existing façade.

Left: Roof top.

Opposite: Entire building, with Second Street in foreground, Pacific in background.

Photography: Tim Street Porter.

It doesn't take a huge project to demonstrate cutting-edge environmental design. This distinctive but unassuming 13,800-square-foot building exemplifies it. Here the client – a nonprofit dedicated to protecting the environment – challenged the architects to create one of America's first LEED Platinum structures, and the resulting building achieves that standard. The first step in making the building environment-friendly was locating it within the pedestrian-oriented downtown Santa Monica, convenient to the city's "park once" garages (once for all downtown destinations), which reduce local traffic and minimize total parking spaces required. NRDC also provides bike racks and showers for bike commuters. The building itself is a reuse of an existing structure, and it generates no carbon dioxide emissions at all. Three light wells direct daylight deep into the interior, and operable windows have Low E double glazing. Sensors control artificial lighting, dimming it in shared spaces as daylight increases, turning it off in unoccupied offices. Photovoltaic cells installed on the roof fill 20 percent of the structure's power needs and, when consumption is low, return power to the grid. For the rest of its energy needs, NRDC buys renewable energy generation credits (wind certificates), so that the building is 100 percent powered from renewable sources with no carbon emissions. Insulation that exceeds code requirement (R-27 for the roof, R-19 for the walls) reduces air conditioning needs – hence power required and release of heat to the outside. To the extent air conditioning is required, it is provided by a high-efficiency displacement system that focuses cool air where it is needed. Heating and cooling for any individual office shuts off automatically when a window is opened. With rain water harvested for gray water use and waterless urinals, the building uses 60 percent less water than conventional office buildings, and it collects 100 percent of rainwater falling on its roof for reuse. All site surfaces outside the building have porous surfaces that pass moisture to the soil below. Over 90 percent of the building materials are either recycled or recyclable. More than half of the materials were either harvested or manufactured within 500 miles of the site. Seventy-five percent of the steel is recycled. Over 92 percent of deconstruction materials and construction waste were recycled. On the building's street floor, a central entrance leads to the David Family Environmental Action

Elizabeth Moule & Stefanos Polyzoides
Architects and Urbanists

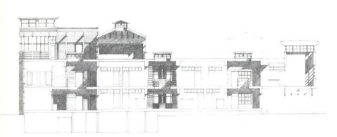

Center, a public amenity that helps maintain activity along the street. Exhibits here include real-time monitors of the building's sustainable systems. A side portal leads to NRDC's offices. At the center of the second floor is a conference room, with clerestory lighting from one of the three "lighthouses" along the building's axis. Appearing over shared spaces, these lighthouses are both functional, in terms of lighting and ventilation, and metaphorical, representing beacons and evoking the nearby ocean. The Hardiplank sustainable wood substitute used on the exterior appears inside in the light wells, and other environmentally friendly interior materials include bamboo and linoleum flooring, carpet made from recycled nylon, panels made from wheat straw, and partitions made of recycled water bottles. Furniture includes custom items made from recycled wood and chairs with webbing made from recycled seat belts. A commissioning team was involved to ensure that sustainable design intentions were carried out during construction and that the occupied building is operated as intended.

Top left: Rear-to-front section.

Above: Rear wall, related in scale and details to nearby houses.

Far left: Leonardo DiCaprio e-Activism Zone in first-floor David Family Environmental Action Center.

Left: Roofdeck with Pacific view.

Opposite: Reception lobby and first floor light well.

Photography: Tim Street Porter.

Elizabeth Moule & Stefanos Polyzoides
Architects and Urbanists

Del Mar Station
Pasadena, California

Right: Aerial perspective of main plaza, with restored station at left, passage to Arroyo Boulevard under pool deck at right.

Below: Transit line passing through residential building to mid-block station.

Below right: View of complex with Arroyo Boulevard in foreground.

Bottom left: Station plaza, with opening to Arroyo Boulevard passage beyond.

Photography: Conrad Lopez, Warren Aerial and Peter VanderWal.

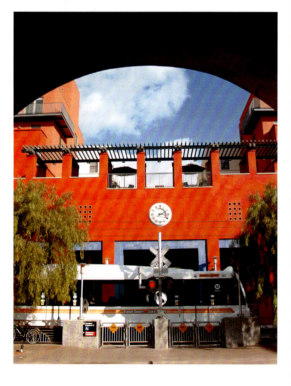

The vitality of downtown Pasadena is greatly enhanced by this project, which assembles 437 residential units around a station on the Gold Line, a light-rail link to downtown Los Angeles. With a total of 433,374 square feet, the complex also includes some neighborhood retail around a new station plaza. The surface transit line bisects the 3.4-acre site and literally runs through the buildings. A zoning variance allowed for buildings varying from four or five stories along perimeter streets to seven stories at the core of the site – rather than the "crew cut" uniform height that could have accommodated the same volume under code. The rail right-of-way is detailed like a public street, with sidewalks, street lights, and residential stoops. An axial passage from Arroyo Boulevard emerges at the station plaza from under the project's pool and recreation deck. Each of the project's four residential quadrants is wrapped around its own courtyard. Four-story portions facing existing industrial buildings across Arroyo Parkway include street-level live-work units, which will encourage pedestrian activity. Other buildings range from Modern to Mediterranean in style, with varied materials, colors, window and door details, balconies, and roofs. In support of environmental goals, there are operable windows in all units and 600 parking spaces for transit commuters. The project includes the adaptive reuse of the 1920s railroad station, which was moved off-site to allow construction of subterranean parking and will be returned and restored to house a restaurant.

Elizabeth Moule & Stefanos Polyzoides
Architects and Urbanists

Rio Nuevo
Tucson, Arizona

This 13-acre project lies within the 62-acre Rio Nuevo project, an area leveled for 1960s urban renewal that is being redeveloped as part of a municipal strategy to reinvigorate the downtown. The plan calls for 150 dwellings in multi-family, mixed-use courtyard buildings, 110 patio-type single-family houses, and 1.54 acres of plazas and green spaces. On the east edge of the development, a Mercado designed in the Southern European and Central American tradition will provide a neighborhood focus, and along the south edge will be two museums serving the entire city. The street layout is influenced by traces of 5,000-year-old Hohokam canals, which lie just below the surface. A variety of street types, all narrower than otherwise allowed by code, is supplemented by alleys and pedestrian ways. Reflecting the long history of desert urbanism, public spaces are shaped to be shaded by abutting buildings and the trees that will flourish in sheltered places. A form-based code calls for two zones, six frontage types, five architectural types, and four styles – the latter adaptations of vernacular types to current building conditions. Construction is to be largely of rammed earth and adobe. Four local builders involved in the project have established reputations for high-quality construction with traditional materials and techniques.

Top left: Illustrative plan.
Top right: New adobe and rammed earth houses.
Above: Pedestrian paseo.
Above Right: Avenida del Convento with Mercado at right.

MSI

462 S. Ludlow Alley	572 E. Green Street	934 Lake Baldwin Lane
Columbus, OH 43215	Suite 200	Orlando, FL 32814
614.621.2796	Pasadena, CA 91101	407.896.4007
614.621.3604 (Fax)	626.304.0031	407.896.5677 (Fax)
www.msidesign.com	626.304.0857 (Fax)	

MSI

Crocker Park
Westlake, Ohio

Having developed as a commuter community without a true town center, the city of Westlake saw this project as an opportunity to establish an identity and a gathering place for residents, at the same time filling a need for retail in a region that was underserved. Because of the large scale and novel nature of the project for the city, an extensive multi-year zoning process was undertaken, culminating in approval for a true mix of uses, with retail, civic, and residential elements sharing the same buildings. Key to the project was creation of an engaging pedestrian environment while accommodating the suburban auto-dependent patterns of arrival and departure. Parking garages are buried behind structures, with strong urban elements along the primary public corridors. The project area of about 60 acres includes a one-acre central park, various smaller parks and gathering places at the theater, and alley connections. A chess garden and a seasonal skating rink help to make the development's open spaces regional destinations, as well as amenities for its residents. The 994,000 square feet of buildings include about 40 retail stores, 10 restaurants, and 220 residential units.

Top left: Site plan, showing mixed use development in darkest tone, parking structures in lightest.

Above left: Shopping juxtaposed to lush greenery.

Above: Mixed-use street, showing brick and specialty-finish concrete sidewalks.

Photography: Ellen Puckett Photography, Bialosky & Partners, MSI.

Far left: Urban intensity at an intersection.

Left: Spot for pedestrian relaxation.

MSI

Celebration Hotel
Celebration, Florida

Left: Hotel and other town buildings seen from lake.

Middle left: Hotel seen along public street, with entrance motor court.

Bottom left: Lakefront promenade.

Photography: MSI.

This 115-room boutique hotel overlooks a pristine lake in the neo-traditional setting of the planned community of Celebration. Designed by Graham Gund Architects, the hotel occupies most of its 101,000-square-foot site. MSI's mission was to create appealing outdoor spaces on the remaining 35,000 square feet, including an arrival motor court, a pool and spa, gardens, walkways, and seating areas, all integrated with the town's open space patterns. In line with the Central Florida setting, subtropical elements such as palm trees and concrete paving with coquina shell aggregate are mixed with selections from slightly more northerly climates, such as brick paving and live oaks.

MSI

North Bank Park
Columbus, Ohio

About 14 acres of brownfield riverfront land north of downtown Columbus has been transformed into a park with walkways, bike paths, a park pavilion, and a children's fountain plaza. The new park makes a key downtown link for recreational users and bicycle commuters to a trail that extends over ten miles. Previously occupied by industries and a railroad embankment, the site posed environmental challenges. Abandoned fuel tanks had to be removed, along with concrete footing and slabs, but no contaminated soils were unearthed. Because of highly compressible soils, about 220 piles driven 60 feet deep were required to support new buildings and retaining walls. Removal of the existing embankment prompted a requirement that the park provide flood protection at modern levee standards. Extensive negotiation with federal and state agencies was required to permit narrowing roadways and introducing on-street parking along them. Both sides of a 1,340-long retaining wall in the park are faced with limestone salvaged from the Ohio Penitentiary, which had occupied an adjacent site. Other recycled material includes stabilized crushed brick and sand paving. Native and naturalized tree species and drought- and heat-tolerant fescue grass have been planted throughout the park.

Top left: Swath of park with railroad bridge and downtown towers beyond.

Top right: Fountain and glazed pavilion on park's central terrace.

Right: View with riverfront park in foreground, downtown Columbus beyond.

Bottom right: Riverfront walkway.

Photography: MSI, Feinknopf

MSI

Nationwide Arena District
Columbus, Ohio

Left: Arena with plaza at approach from downtown core.

Above: Quiet area in central park.

Bottom left: Plan of district, with arena large block at upper right, wedge-shaped park at lower right, ballpark and residential area at left, North Bank Park and river at bottom.

Opposite: Dining and entertainment portion of district.

Photography: Ellen Puckett Photography, MSI, Nationwide Realty Investors.

The construction of the Nationwide Arena is just one element in the redevelopment of a 79-acre neighborhood in downtown Columbus. Once the site of the Ohio Penitentiary, along with various industrial uses, the area is being transformed into a walkable environment for residents, workers, and visitors. Considerable environmental remediation and infrastructure improvement was needed. The plan integrates the volume of the new 19,000-seat arena into the community and provides pedestrian-scaled plazas at its entrances, which are appealing public amenities at all times. The arena building was designed by 360° Architecture, known when it was designed as Heinlein Schrock, with planning and site work by MSI, as in the entire district. Necessary parking is provided in lots close to the sports and entertainment venues and in several garages located inside development blocks in order to maintain the continuity of active street frontage. Besides the arena,

MSI

Above: Characteristic pedestrian alleys in district.

Right: Arch from city's demolished Union Station – sometimes called Burnham Arch for its architect – at entrance to central park.

Bottom right: Central three-acre wedge-shaped park.

Photography: MSI, Ellen Puckett Photography.

programmed uses include about 1.5 million square feet of office, retail, and entertainment facilities, about 450 residential units, a 1,700-seat movie theater, a 4,000-seat live concert venue, and an 11,000-seat AAA baseball stadium.

Reestablishing a street grid for the area required reconciling two differently oriented street patterns, and this was resolved by creating a three-acre wedge-shaped park, a central public space for the neighborhood. Installed at the narrow end of the park, adjacent to the arena, is an arch and other architectural fragments from the city's Union Station, demolished in the 1970s. At the broad end, the park faces the new North Bank riverfront park (the preceding project on these pages), which borders much of the district. Connections to the downtown core are reinforced by pedestrian links over and under the railroad tracks along one edge of the site. Construction of new brick-paved streets, the first to be built in the city for many decades, required extensive coordination with city engineering officials.

MulvannyG2 Architecture

1110 112th Avenue NE	601 SW Second Avenue	18200 Von Karman Avenue	8484 Westpark Drive	NanJinXi Road 1038#
Suite 500	Suite 1200	Suite 910	Suite 700	The West Gate Mall, Ste. 2001
Bellevue, WA 98004	Portland, OR 97204	Irvine, CA 92612	McLean, VA 22102	Shanghai, CH
435.463.2000	503.223.8030	949.417.9707	703.564.8484	011.86.21.62185761
435.463.2002 (Fax)	503.223.8381 (Fax)	949.417.9708 (Fax)	703.564.8400 (Fax)	011.86.21.62178525 (Fax)
www.MulvannyG2.com				www.MulvannyG2.com/zh-cn

MulvannyG2 Architecture

Redmond City Hall
Redmond, Washington

The City Hall was commissioned through a competition in which three local finalists submitted designs. The unanimous choice of the jury and the City Council, this scheme provides efficient office space in a structure with a public presence appropriate to the home town of Microsoft and Nintendo. Offices with ample natural light are located in east and west wings, connected by an atrium with glazed front and back walls. A broad entrance canopy on tall columns responds to the regional climate and architectural traditions. Partly sheltered under it is the drum of the Council Chamber, clad in aluminum panels hand-painted in a copper color, which is less expensive than copper and does not change color. The atrium and office wings overlook a landscaped terrace with views of the Sammanish River. Sustainable features include: recycling of construction debris; use of recycled and local materials; water-efficient irrigation and drought-resistant landscaping that reduce water demand 50 percent; bike storage and changing rooms; electric vehicle recharging stations; reflective roofing and covered parking to reduce the heat island effect; use of daylight illumination. The city is pursuing LEED Silver certification. The 113,000-square-foot structure houses 260 employees, with room inside for expansion.

Top left: Building seen from main approach.

Top right: Terrace between wings, with back entry to atrium.

Left: Council Chamber rising from reflecting pool with sculpture by Ed Carpenter.

Bottom: Concept sketch.

Opposite: Civic-scaled and climatically appropriate entrance canopy.

Photography: Image © Pisano 2006, this page; Image © Steve Keating 2006, opposite page.

MulvannyG2 Architecture

Zhejiang Fortune Financial Center
Hangzhou, China

Far left: Early concept sketch.
Left: Model of complex.
Bottom: Entry plaza with gardens and riverside walks.
Opposite: Office towers joined by retail podium.
Renderings: MulvannyG2 Architecture.

An international design competition entry, this project would be the central landmark of Qianjiang new city in Hangzhou, the capital of Zhejiang province. It consists of a 53-story office tower, a 24-story office tower, and a three-story podium for retail and commercial uses. An axis running through the podium connects the city's government center with a bus terminal and a future subway station. Three underground floors, to be connected to the subway, include a supermarket, maintenance facilities, and a parking garage for 461 automobiles and 3,575 bicycles. The main tower displays a curved curtain wall wrapped part way around a tapered hexagonal volume. On the lower tower, a curved wall with horizontal window bands is situated as if to embrace the larger structure. Outdoor spaces on the 187,030-square-foot site include a terraced sunken plaza for cultural performances and everyday use. An ecological garden exhibits species of the region, and an interpretive trail winds along the adjoining waterway. The total building area of the complex is 1,916,000 square feet, with 983,100 in the taller tower, 277,800 in the lower one, and 214,000 in the podium.

MulvannyG2 Architecture Zhangjiang Semiconductor Research Park, Phase II
Shanghai, China

Phase II of the Semiconductor Research Park will provide 2,316,400 square feet of floor area and is a key component of a larger high-tech park, helping to establish Shanghai as a technology center. The design concept is meant to encourage start-up and growth, with small-office-type environments. The buildings are composed of highly flexible modular blocks of various sizes, which retain their integrity when combined in different configurations. Amenities provided in the buildings include top-floor lounges, meeting spaces, and roof gardens. A consistent vocabulary of curtain walls, overhangs, and entrance canopies maintains the development's identity throughout. The 118-acre site is treated as a true park. Each building has its own garden and is connected to others by landscape or water links. A central public activities area includes a water feature and a landmark tower.

Left: Park viewed from across the Lu Jia Bang River.
Bottom: Small-scaled buildings facing landscaped open space.
Opposite top: Aerial view of entire Phase II.
Opposite bottom: Building volumes with characteristic details.
Photography: Sheng Zhonghai.

MulvannyG2 Architecture

Fudan Crowne Plaza Hotel
Shanghai, China

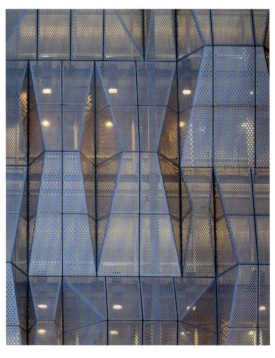

Situated opposite the prestigious Fudan University, the hotel is within easy reach of Shanghai's central business district, the recently developed Pudong area, and major tourist attractions. The 20-story structure was designed to fit harmoniously into its surroundings. Designed as five-star property, the building includes 309 rooms and 34 suites, all in contemporary style, plus restaurants, a multifunction hall, conference facilities, a swimming pool, and a fitness center. The top floor houses a President's Club with additional amenities. The structure's north-south orientation conserves energy, and the lobby's glass roof captures abundant natural light. A central courtyard and surrounding gardens help to blend nature and the built environment.

Top left and right: Façade with metal screen forming metaphorical waterfall.

Above: Main lobby.

Left: Detail of metaphorical waterfall wall.

Photography: Sheng Zhonghai.

NEWMAN GARRISON GILMOUR+PARTNERS

20401 SW Birch Street
Suite 200
Newport Beach, CA 92660
949.756.0818
949.756.0817 (Fax)
www.NGGPARTNERS.com

NEWMAN GARRISON GILMOUR+PARTNERS   Watermarke
Irvine, California

Faced with a very competitive residential market, where Mediterranean architectural themes predominate, the decision was made to employ a more formal, urban, and cosmopolitan – inherently less suburban – approach in this project. Inspired by Northern European traditions, the architecture here is targeted toward the upscale buyer, with high levels of finish and detail carried through interiors, exteriors, and landscaping. The site's high water table ruled out a parking-podium solution, so parking consists of two multi-story above grade garages with residential units wrapped around them. Local codes required several open spaces of ¼ to ⅓ acres each, with provisions for active recreation in some of them, plus multiple fire lanes for access to the wood-framed structures. The ample landscaping of fire lanes and the orientation of courtyards toward mountain views assure pleasing outlooks from units. Ten distinct layouts for the development's 535 condominium units include townhouses, flats, and larger, more elegantly appointed penthouses. The two-story, 8,000-square-foot clubhouse includes a multimedia entertainment room, a surround-sound screening room, a library with fireplace, a fully equipped conference center, a great room with bar, and a chef's kitchen. Adjoining it is a two-story sports area, with tennis and basketball courts. The Junior Olympic pool features a spa, and the staffed, well-equipped fitness center occupies its own building. Elsewhere on the site are two satellite pools, four spas, volleyball courts, a playground, a meditation garden, and walking trails.

Above: Site plan.
Below left: Units surrounding axial landscaped court.
Below: Luxuriant plantings seen from windows and balconies.
Opposite: Classical design motif and predominantly deciduous trees reminiscent of Northern Europe.

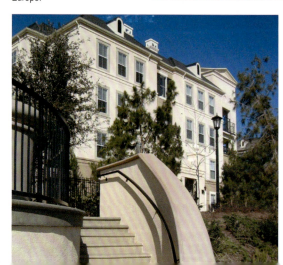

Right and far right: Clubhouse with central pool and related amenities.

Photography: Will Hare Photography.

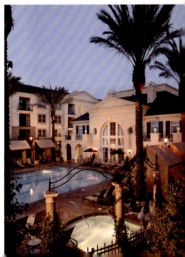

NEWMAN GARRISON GILMOUR+PARTNERS   Avalon Del Rey
Los Angeles, California

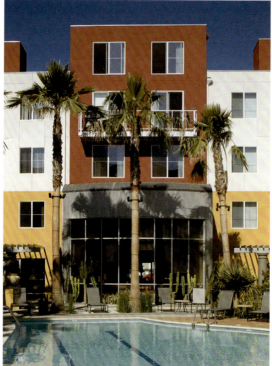

Located in a transitional industrial location, this development provides its predominantly single and professional residents with top-of-the-market on-site technology and recreational amenities. On a 4.5-acre site in the Playa del Rey master-planned community, the project includes 309 units – including stacked flats and lofts—in two four-story courtyard structures. Since the site was subject to a height limit as measured from the lowest point of adjoining sidewalk, the building pad was depressed about 2 feet below the sidewalk for a portion of the project. Recognizing the area's office and light industrial uses, the buildings were given a tech-industrial design treatment, with relatively edgy material such as corrugated metal siding, plus bold colors to emphasize the cubic break-up of their masses. Parking at a ratio of 2:1 is provided in two five-level, above-grade structures, which are sited to provide light and sound buffers between this project and adjoining commercial properties. On-site amenities include a central pool and a 4,903-square-foot clubhouse.

Above: Street front, showing mixed residential-industrial imagery.
Top right: Multicolored facades overlooking central pool.
Above right: Club room overlooking pool.
Right: Main lobby.
Opposite: Residential structures surrounding courtyard.
Photography: Steve Hinds Photography.

Newman Garrison Gilmour+Partners

Paxton Walk
Las Vegas, Nevada

The generous mix of uses on this 29.4-acre site required extensive negotiation with city staff and adjacent homeowners. The original zoning called for a maximum of three stories, but the developers were able to get approval for four stories, mainly along the Centennial Parkway edge – if they provided well articulated building volumes with well landscaped setbacks – so that occupants can enjoy views of downtown Las Vegas. Also approved were two-story townhouses along the back side of the property if there was a substantial landscaped buffer bordering the adjoining large-lot single-family developments. The project's 1,011,920 square feet of buildings are laid out as four "villages" and a town center. The program includes 782 residential units, 33,750 square feet of retail, 57,650 square feet of offices, and 14,289 square feet of restaurant. The unit types range from family-oriented townhouses to lofts and flats for singles and couples. The inward-oriented building blocks create intimate, shaded courtyards for shoppers. Planted streets encourage pedestrian circulation throughout the project. The European Classical design theme lends a consistent identity to residential and commercial portions of the development, and the bulk of the structures is minimized by the division of their facades, with a variety of surfaces, roofs, openings, and balconies.

Opposite top: Rendering of town center, showing variety of surfaces, openings, roof, and roodlines.

Opposite bottom: Generously planted courtyard setting for retail and restaurants.

Top: Overall plan, with four "villages" flanking town center, townhouse rows at top.

Above: Model of town center surrounding central courtyard.

NEWMAN GARRISON GILMOUR+PARTNERS   Amerige Pointe
Fullerton, California

For a site adjacent to commercial properties, this residential project includes commercial uses near the existing retail. Since the commercial market here was difficult to gauge, some of the commercial street frontage is adaptable to work/live units. Angled parking in front of commercial and work/live units is supplemented by residential and employee parking at the rear of buildings. The architectural treatment, recalling commercial and warehouse architecture of the early 1900s, provides a variety of massing and points of interest. Loft units have expanses of glass and wrought iron balcony railings. A semi-circular restaurant space at one corner acts as a focal feature. A consistent architectural approach maintains the identity of the project through the transition to the three-story garden apartments occupying most of the site. Along the major street frontage, the apartments continue the building line of the mixed-use buildings. In the core of the development are generous landscaped areas with amenities shared by residents.

Above: Street front by night, showing angled parking in front of street-level retail.

Top left: Mix of commercial and residential units facing business-oriented street.

Left: Site plan, showing mixed commercial/residential buildings along streets at lower left, garden apartments on rest of site.

Bottom: Conceptual streetscene elevation.

Photography: Steve Hinds Photography.

O'Brien & Associates

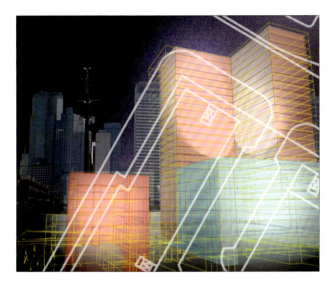

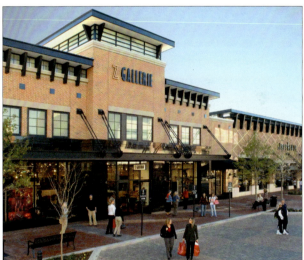

5310 Harvest Hill Road #136
Dallas, TX 75230
972.788.1010
972.788.4828 (Fax)
www.obrienarch.com

O'Brien & Associates

City Lights
Dallas, Texas

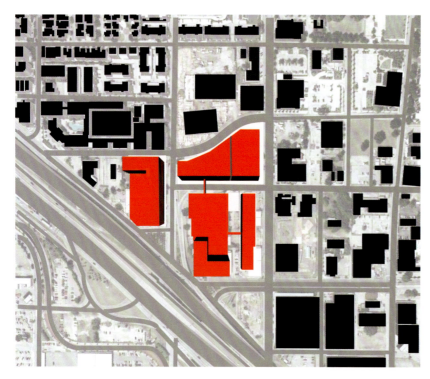

Located on the east side of Central Expressway, adjacent to downtown Dallas, City Lights is a mixed-use project containing 120,000 square feet of retail, a 41,000-square-foot fitness club, 770 apartment units, and 2,200 parking spaces. An urban street grid divides the project into three parcels, with shops fronting the internal streets. Three apartment towers with wide views of Dallas rise from the retail parking garages, which are easily accessible from the street. Residential parking is below grade. The large scale of the complex is broken up by offset façades, implying separate buildings erected over a period of time. The look is that of New York's Lower East Side, with a variety of brick and cast stone facades, metal awnings, fire escapes, and rooftop water tanks. Cafes and dining patios are located along the main street.

Left: Site and location plan.

Bottom left: Artist's interpretation with downtown Dallas in background

Bottom: Facades broken up by offsets, textures, and colors to imply multiple buildings.

Opposite: Exploded axonometric.

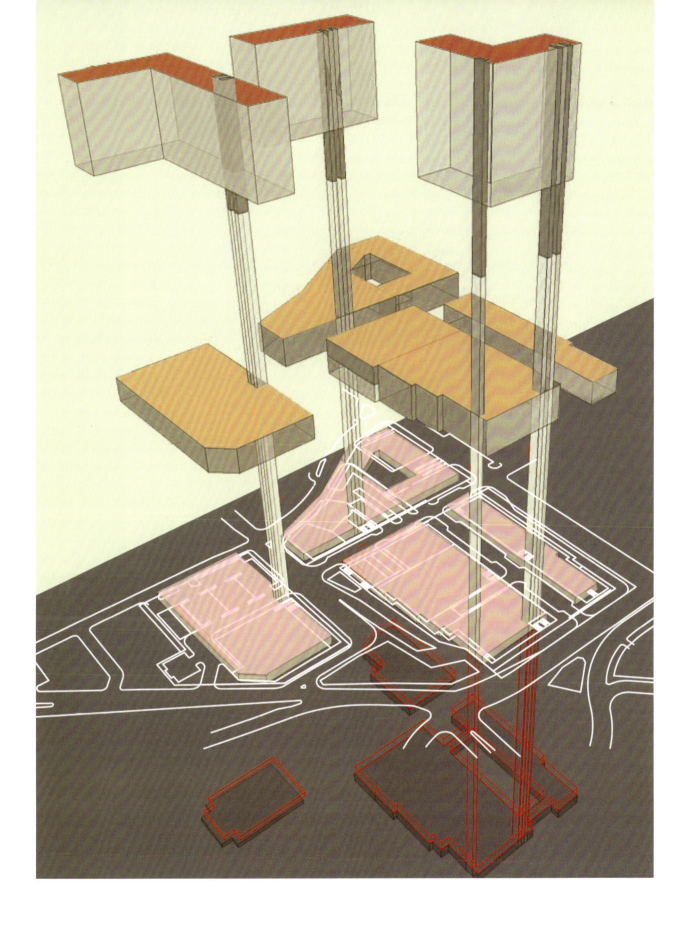

O'Brien & Associates

The Harbor
Rockwall, Texas

Right: Center fountain.

Below right: Original site plan by Good Fulton & Farrell Architects.

Bottom right: Color details of the restaurant building.

Opposite top: Tower under construction and lake.

Opposite bottom: Mediterranean-style building fronts.

The Harbor is a 273,500-square-foot high-end specialty center conceived as a Mediterranean village on the bank of Lake Ray Hubbard. The City of Rockwall developed an 11-acre waterfront park between the center buildings and the lake. The park and buildings were seamlessly integrated by the designs of TBG, who was commissioned by both clients. An arm of the lake reaches into the center of the project and terminates in a large fountain court, around which are located a theater, restaurants, and shops. The drop in grade toward the lake is handled by terracing the parking lots and establishing a difference in grade between the parking side and the lake side of retail buildings. Some buildings have second-floor office space facing the lake. Numerous passageways connect the parking areas to the park, penetrating the shops that stretch from the theater to the hotel. Materials include tilt-up walls painted in earth tones, with accents of stone and EIFS, and concrete tile roofs.

O'Brien & Associates

The Shops at Highland Village
Highland Village, Texas

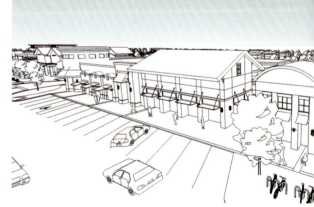

Left: Conceptual plan of development.

Opposite top: Bird's-eye view of retail, theater, and office development.

Opposite bottom left: View showing sidewalks and on-street parking.

Opposite bottom right: View indicating variety of building design.

This 378,000-square-foot mixed-use development includes a major bookstore, a theater, and numerous specialty shops. The plan is a grid similar to a small downtown, with buildings differentiated in scale, design, and color, so that each tenant occupies a distinctive property. As the project morphed from a big-box development to a high-end center, the city was pleased with the "urban grid" planning approach. The bookstore at the center has entries to the north and south. The north side faces an active public space with a pop fountain, surrounded by restaurants and a theatre. The South side of the bookstore opens toward a more formal space with a reflecting pool and a way-finding obelisk. There are 32,500 square feet of offices above stores along this formal plaza. The perimeter of the development accommodates a supermarket and several freestanding restaurants. The entire 40-acre site includes a 2.9-acre retention pond and 3.8 acres of trails.

O'Brien & Associates

Sugarland Town Center Expansion
Sugarland, Texas

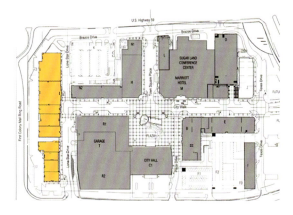

Left: The Z Gallerie as a focal point at end of the main street.

Middle left: Elevation of new retail buildings.

Bottom left: Site plan, with new retail buildings in yellow.

Photography: Larry Pullen.

This project consists of two buildings totaling 50,000 square feet at one end of the Sugarland Town Center's already successful main street. While the existing buildings had been homogeneously designed, the center's owner wanted the architects to provide more design diversity in the new structures. The added buildings have a variety of cornices and bases of polished black granite, limestone, or poured concrete with an EIFS finish. Three different face bricks are used on the walls, with various patterns and bonds. At the center of the new line of buildings, the Z Gallerie provides an inviting focal point at the end of the main street.

Perkowitz+Ruth Architects
Studio One Eleven at Perkowitz+Ruth Architects

Corporate Headquarters:
111 West Ocean Boulevard
21st Floor
Long Beach, CA 90802
562.628.8000
562.628.8005 (Fax)
www.prarchitects.com

Regional Offices:
Orange County, CA
Portland, OR
Las Vegas, NV
NW Arkansas
Washington, DC

Studio One Eleven

Court Street West Specific Plan
San Bernardino, California

Left: Proposed site plan, showing reused buildings from former mall.

Middle left: Public green at junction of commercial and residential areas.

Bottom left: Entrance to site through residential area.

For this redevelopment of a mall site, a street grid has been reintroduced, with smaller than usual blocks to produce a dense, pedestrian-friendly, mixed-use district. Located within one quarter mile of a transportation node, the development of active street frontages will further encourage pedestrian activity and reduce automobile use. Four buildings from the original mall will be retained and reused for retail and offices. A large public green and several smaller semi-public green spaces will help with the management of storm water and reduce the amount of impervious surface. Provisions for shared parking among different uses have lowered parking requirements.

Perkowitz+Ruth Architects

Galerias Hipodromo
Tijuana, Mexico

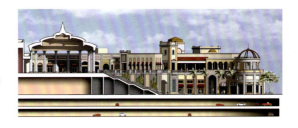

The 345,000-square-foot retail and entertainment complex features 35 local shops, restaurants, cafés, and a 12-screen movie theater around a big box anchor tenant. The prime tenant's necessary large volume has been laminated with shops and topped with a cinema. Two levels of parking underlie the entire development. At the heart of the project is the Main Plaza, with a performance stage that has been incorporated into the design of a water feature. Kiosks, shops and cafés define the circular open space. A grand stair leads up to a colonnaded rotunda outside the cinema on the second level where terraces overlook the Main Plaza. Sun control has been integrated throughout the complex in covered walkways, colonnades and awnings.

Top left: Stair and second-level terraces overlooking plaza.

Top right: Section showing layers of retail above two underground parking levels.

Left: Plan showing circular layout around Main Plaza.

Bottom: Overall view of development.

219

Studio One Eleven

Downtown Long Beach Visioning
Long Beach, California

Downtown Long Beach, comprising 150 city blocks, is currently under strong development pressure for high-density residential mixed-use projects. This study is a basis for updating the city's downtown zoning and will be integrated into its General Plan. In a short time, the study generated an understanding of the existing urban context through a three-dimensional digital model incorporating building location and height, streets, open space, light rail routes, and designated historic structures and districts. Animated fly-throughs were developed as tools for community outreach and education. Concurrently, the study analyzed high-rise development standards in cities with similar growth patterns, including Vancouver, San Francisco, Portland, Seattle and San Diego. Preliminary standards covering massing, protection of view corridors, solar access, and sea breeze patterns were drawn up and tested on a dozen proposed developments to see what changes they would require. Working with key community stakeholders and local officials, the firm continues to refine the standards. Major strategies for a pedestrian-friendly environment include greater land-use intensity, reduction of surface parking lots, shared parking facilities, and support of alternative transportation means, including light rail, shuttles, buses and bikes. A massive tree-planting program is proposed to beautify the downtown area and reduce its heat-island effect. Habitat improvement and greater pedestrian access are proposed for natural resources adjacent to downtown, including the Los Angeles River and Pacific Ocean shoreline.

Top left: Open space plan, with park areas in shades of green, hardscape in tan.

Top right: Diagrammatic view of city core identifying transit routes. Scenario promotes building intensity along key corridors.

Above: Plan showing neighborhoods.

Perkowitz+Ruth Architects

Bella Terra
Huntington Beach, California

This project successfully repositioned the formerly enclosed Huntington Beach Mall, which had become dilapidated and unleasable. The $170-million revitalization transformed it into a 770,000-square-foot open-air shopping, dining and entertainment destination with an active pedestrian environment. Three existing major tenant buildings that were retained became the armature of the new plan. Medium-format retailers and specialty shops were layered around these structures, softening the scale of the large volumes by breaking up their abrupt planes and giving all tenants exterior exposure. Pedestrian paseos radiate from a central amphitheater. One end of the complex is anchored by a 20-screen cinema and the amphitheater. At the opposite end, a pond and courtyard encourage relaxation. The project's limited street-front visibility was addressed by boldly facing portions of it toward the neighboring freeway and using vibrant colors to attract attention from adjacent boulevards.

Top: Central amphitheater and one of the radiating paseos.
Right: Identifying campanile and dining terrace.
Bottom right: Pond courtyard.
Photography: Paul Turang Photography, Lawrence Anderson.

Studio One Eleven

The Garden
Taichung, Taiwan

On a 5.4-acre site in a dense neighborhood, The Garden's three residential towers accommodate 2 million square feet of floor area. Two 18-story towers flanking a serenely landscaped entry court frame the 26-story central tower, which is topped by a spacious shared lounge offering expansive views. Elegantly articulated facades are based on fenestration that expresses the "public" and "private" spaces within the units. The "public" spaces such as living rooms have floor-to-ceiling window walls with vertical mullions, offering the least obstruction of views. Façades of the "private" spaces such as bedrooms feature balconies, overhangs and decks that provide visual screening and outdoor areas. Energy efficiency is accomplished through several devices: shading of exteriors with balconies and overhangs; use of louvers for sun control; use of translucent channel glazing to admit maximum daylight; provision of operable windows; use of green roofs and light-colored roofing; integrating skylights into the landscaping of the central court to provide daylight to the services and parking below; use of light-colored, high-insulation-value, durable exterior materials (i.e. natural stone, "Inax" tile, pebble-finished concrete) which require no painting and little maintenance.

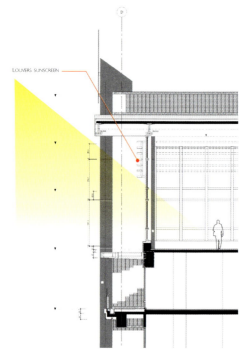

Top left: Aerial view of project's three towers.
Top right: Three-tower complex as seen from street.
Right: Characterisic exterior wall section.

Studio One Eleven

Melrose Triangle
West Hollywood, California

Left: Entire triangle, penetrated by paseos, with distinctive Gateway Building at apex marking entry to West Hollywood.

Below: Large-scale façade along Santa Monica Boulevard.

Bottom: Entrance from Melrose Avenue to through-block paseo.

Studio One Eleven has provided input for the city of West Hollywood on its development of a municipal Green Building program. This project demonstrates how the city's intended goals can be implemented in a large-scale, mixed-use, privately financed development, addressing not only technological strategies, but the larger urban context. Located at the key intersection of Santa Monica Boulevard, Melrose Avenue and Almont Avenue, the project accommodates 200 residential units, including lofts, flats and townhouses, along with 78,000 square feet of ground-floor retail over subterranean parking. Extensive community outreach and coordination with city officials ensures the project's compatibility with its community. Its buildings vary in massing and style, with a large-scale façade reinforcing the character of the boulevard, while the avenue frontages are lower and smaller in scale. Open space design has been carefully adjusted to create paseos, plazas and courtyards that contribute to the development's ambience and invite pedestrians to pass through. Tenants such as fashion boutiques, restaurants, showrooms and furniture and art galleries will be selected for their ability to help re-brand the district as pedestrian-friendly. Sustainability features include overhangs, light shelves, low-e glass, native plantings to reduce irrigation demand, and resource-efficient materials such as fiber cement siding, integral stucco, resin-based phenolic cladding, and other materials with recycled content.

Perkowitz+Ruth Architects

Bridge Street Town Centre
McKinney, Texas

Located in one of the nation's fastest-growing cities, this development is planned to accommodate about one million square feet of upscale commercial space. Included will be 233,650 square feet of retail, 132,000 square feet of offices, 69,200 square feet of restaurants, an 84,000-square foot theatre, and a 210-key, 550,000-square foot hotel and convention center. With buildings of European character, combined with over four acres of green open spaces, the project's lake, waterfalls, rapids and footbridges will lend the project distinction. Sun control has been integrated into the complex through covered walkways that shade the storefronts and enhance the walkable environment.

Top: Elevation of theater building.
Middle: Theater and flanking shops seen across fountain plaza.
Bottom: Hotel viewed from terrace overlooking lake.

RLC Architects, P.A.
Retzsch Lanao Caycedo

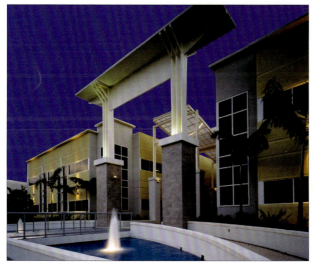

137 West Royal Palm Road
Boca Raton, FL 33432
561.393.6555
561.395.0007 (Fax)
www.rlcarchitects.com

RLC Architects, P.A.
Retzsch Lanao Caycedo

Aqua Vista Lofts
Fort Lauderdale, Florida

Above: View including three of development's six rows of houses.
Left: Site Plan.
Bottom left: Interior showing concrete floors, exposed structural members.
Opposite: Townhouse fronts.
Photography: Chuck Wilkins Photography.

Located on the outskirts of the urban core, the project adapts the urban grid to an irregular site. Vehicular traffic can flow between the two parallel thoroughfares bordering the site, and access to individual garages is through rear alleys. The house fronts are all reached along landscaped pedestrian ways. Limited to 25 units per acre, the plan organizes 31 units on the 1.32-acre site. With an intended market of young professionals, the townhouses adopt a loft style, with exposed construction materials, high ceilings, and stainless steel kitchen packages. Glazing is extensive in living areas, limited in more private spaces. Each three-story townhouse includes a two-car garage, living room, kitchen, three bedrooms, and three-and-a-half baths. Roof terraces with hot tubs provide the amenity of outdoor private space with views of the urban surroundings.

RLC Architects, P.A.
Retzsch Lanao Caycedo

U.S. Epperson/Lynn Insurance Group Corporate Headquarters
Boca Raton, Florida

A sweeping curve of polished black granite behind a front wall of stainless steel and glass establishes the distinctive identity of this corporate headquarters. An 85,000-square-foot building replacing the client's previous office building on the same property, the design embodies sleek Modernism, rather than Boca Raton's familiar Mediterranean-inspired imagery. The other sides of the building feature white granite columns, full-height glass, and aluminum sunscreens to filter the intense Florida daylight. At one corner, the executive suite is expressed by an all-glass plane, tilted and cantilevered from the

structural piers to distinguish it from the main building volume. Upturned concrete beams allow this expanse of curtain wall to be uninterrupted by columns. In response to the client's desire for an exceptionally quiet working environment, no mechanical equipment has been located on the roof. High performance glass is used in the curtain walls for its acoustical performances, as well as energy efficiency. The building contains a 2,400-square-foot state-of-the-art auditorium, a 2,870-square-foot fitness center, and a four-car air-conditioned executive garage. Some other parking is sheltered at grade under the building, and landscaped parking areas are formally laid out on the six-acre site.

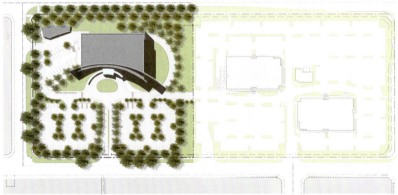

Left: Layered main façade with black granite wall and stainless steel/glass curtainwall. Steel and glass entrance canopy in foreground.

Below: Rendered site plan of Lynn Insurance Campus.

Bottom left: View of front façade from main entrance drive.

Bottom right: Executive board room private balcony.

Opposite: End of polished black granite front wall, with entrance canopy at left, corner executive suite at right.

Photography: Chuck Wilkins Photography.

RLC Architects, P.A.
Retzsch Lanao Caycedo

951 Yamato
Boca Raton, Florida

Left: Renovated building seen from road.

Below: Main entrance before renovation.

Bottom: Entrance area during renovation process.

Opposite: New portal, fountain pools, and curtain walls at main entrance.

Photography: Chuck Wilkins Photography.

At the edge of a commerce park, fronting on a main road, this 153,000-square-foot building was originally support office space for IBM. After it had been vacant for several years, the developer renovated it for the rental office market. Beyond revisions to the building's highly inefficient office layouts and mechanical systems, its lack of identity needed to be addressed. The main façade was altered to admit a better quality of natural light, and the main arrival space given a strong definition. The new façade is designed to be effective at two scales: that of the automobile traffic on the adjacent road and that of the pedestrian entering the building. The main entrance is redefined by an open gallery leading to the interior lobby and atrium. The drop-off area is flanked by two fountain pools separated by a central entry bridge. Natural stone piers support a translucent canopy, and two of them extend upward to form a gateway. The renovation is completed with simple yet refined landscaping and lighting that accentuates key building features.

RLC Architects, P.A.
Retzsch Lanao Caycedo

Domus Office Tower
Hallandale Beach, Florida

Located in Hallandale's core, this 550,000-square-foot mixed-use tower is a major component in the transformation of the area. Sited at the edge of the city's main boulevard facing Gulf Stream Park, it will reinforce the urban character intended for this district. On less than one acre, the complex will include 13,000 square feet of ground-floor retail, parking for 650 vehicles on the next nine floors, 191,000 square feet of office space on the upper 10 stories, and an elevated public plaza at the top to take advantage of magnificent vistas toward the park and ocean. A central core rising through all these components allows for office space with views in all four directions. The architectural treatment relates to the city's prevalent contemporary style, using blue-tinted glass and aluminum curtain walls, pre-cast concrete, and stainless steel mesh. Landscaping at the street level and elevated plazas enhances the experience of the building.

Above: Tower Elevations showing retail, garage and office.
Renderings: Mauricio Villa.

RTKL Associates

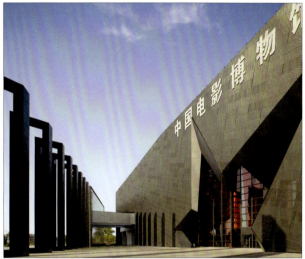

901 South Bond Street
Baltimore, MD 21231
410.537.6000
410.276.2136 (Fax)
www.RTKL.com

RTKL Associates

New Jiang Wan Cultural Center
Shanghai, China

Conceived as the core of an eco-themed development, the cultural center reflects the symbiotic relationship between human behavior and the environment. The New Jiang Wan community is an environmentally sensitive development in Shanghai's last wetland preserve, about 15 kilometers from the city's center. The structure contains 6,000 square meters (64,600 square feet) of exhibition, education, performance, and entertainment facilities on two floors, plus one floor of underground parking. Municipal offices for the administration of the development are seamlessly integrated with the civic amenities of the building, yet independently operated and secured. The central atrium serves as both a public gathering place and a promotional court for the project, and the auditorium accommodates both public and management functions. The building's architecture expresses the constant transformation and symbiosis of natural processes. Rising and diminishing elevations, sweeping curves, and angles are reflected in pools around the building and in the surrounding wetlands. The design team worked with the client and manufacturers to make the structure a demonstration of sustainable methods and inventive applications of green products. The use of composite wood panels on the exterior was a departure for China, approved after work with the vendor to improve its physical properties. The resulting cladding is environmentally sound and extends the warm qualities of natural wood displayed on the center's interior to its outer surfaces.

Top right: Courtyard.

Above: Main entrance after dark, showing translucent glass cladding.

Right: Cultural center as herald of new eco-friendly development.

Opposite top left: Planted roof extending over entrance canopy.

Opposite top right: Detail at entrance.

Photography: Fu Xing Studios.

RTKL Associates

The Chinese Museum of Film
Beijing, China

To recognize the remarkable evolution of the Chinese film industry over the past century, the Beijing Radio, Film, and Television Bureau organized an international design competition for the country's first film museum. RTKL, in a joint venture with the Beijing Institute of Architectural Design and Research, won the commission for the $69-million building, the keystone of a planned entertainment district being developed in anticipation of the 2008 Summer Olympics. The facility accommodates four levels of exhibitions, a cinema complex including an IMAX theater, and research and administrative offices. In its translucent walls and the suspended screens in its ample lobby, the building embodies the basic characteristics of both architecture and film. The museum itself is a simple black box within this more freely organized envelope. Important design consideration was given to flexibility, environmental sensitivity, and establishing the identity of the building and its district outside an established urban context.

Above: Entry plaza with star-shaped portal.

Top: Museum as first building in developing entertainment district.

Middle left: Museum rotunda.

Middle right: Lobby with multiple translucent screens.

Bottom left: Inside star-shaped entrance through translucent front wall.

Bottom right: Stairs and escalator in lobby.

Photography: RTKL Associates.

RTKL Associates

Highmark Data Center
Harrisburg, Pennsylvania

Left: Two levels of building volumes fitting into sloping site.

Below left: Bold precast fin identifying main entrance.

Bottom left: Brick-walled mechanical yard and precast-paneled computer volume seen here to left of employee and visitor portions.

Photography: RTKL Associates.

A medical insurance provider, Highmark wanted to forgo the "gray box" architecture of many data centers in developing a new facility to handle millions of documents securely. Unlike many data centers, this one was also designed to provide a high-quality working environment for some 50 employees working 24/7 and a fitting destination for clients and visitors. It was also designed to meet requirements for LEED Silver certification. The slope of the 11-acre site allowed for a two-story structure with access at grade for both levels. Inside, the 87,000-square-foot facility includes 50,000 square feet of computer space and a 28,000-square-foot data center, wrapped with office and support areas. A ramped glass-lined hallway runs between the command center and data center spaces, providing visitors with a view of both at the same time. Well-designed amenities, including a fitness center, break room, and outdoor gathering area afford employees ample daylight and panoramic views of the landscape. For rapid, economical construction, precast sandwich panels predominate on the exterior, with areas of brick and curtain wall articulating portions of the building. LEED certification required exceptional attention to all aspects of design and construction, since data centers by their nature tend to have high electrical and water demands. The project's green design features include use of rainwater for toilets and cooling towers, under-floor air distribution, alternative vehicle power stations, and recycled and renewable interior materials. By using the RTKL-invented Tower of Cool, the design team was able to reduce overall energy demand by 29 percent, exceeding a minimum goal of 15 percent. Costs of operation will be significantly below those of comparable facilities.

RTKL Associates

Reginald F. Lewis Museum of Maryland African American History and Culture
Baltimore, Maryland

The rich, diverse story of African Americans in Maryland is encapsulated in a 82,000-square-foot structure in downtown Baltimore. Designed by Freelon/RTKL, a joint venture, the museum represents this history and culture in a compact, black-granite-clad volume cleft diagonally by a bright red plane. Because of utility installations under part of the 1.5-acre site – formerly a parking lot – it was not economically feasible to extend the building horizontally. A public plaza links this museum to the nearby Flag Museum. Inside, the museum houses interactive learning areas, meeting rooms, offices, a 200-seat auditorium, an information resources center, a shop, and a café. With much of the street floor devoted to loading docks and other service areas, shop, and café, the main program areas are on the upper floors. A spacious second-floor lobby is reached by a curved stair that passes through the red wall toward increasing natural light as the visitor moves upward. With no established African American style as a precedent, the architects conveyed group spirit, pride, struggle, and accomplishment through a series of symbols and a selective color palette. On the exterior are large-scale quotations and images of African American Marylanders, including Frederick Douglass and Harriet Tubman. Throughout, the colors black, red, yellow, and white recall African traditions, as well as the colors of the distinctive Maryland state flag.

Top left: Overall street view.
Top right: Evening view of entrance front.
Middle left: Atrium organized around red wall.
Middle right: Entrance detail at dusk.
Bottom left: White vestibule and red wall at entrance.
Photography: RTKL Associates.

RTKL Associates

City Crossing
Shenzhen, China

Prior to creation of City Crossing, its site contained an 86,000-square-foot public park, seen by some residents as a place to socialize, by others as an eyesore. The developer agreed to replace the park with more than 80,000 square feet of open area in large and small open spaces in the project. This park network provides multi-level spaces that link the street to below-grade retailers, cafés, and transit connections. Indoor facilities of City Crossing's Phase I include a 1.4-million-square-foot retail center (MIXC), a 452,000-square-foot office building, and 1,000 surface parking spaces. The MIXC, the development's main anchor, balances inward- and outward-facing tenants. Its grand entrance is a three-level retail pavilion with an undulating glass roof. A galleria of high-end retail tenants bridges over a road through the center. A multi-level grand room contains a skating center and seating and dining platforms. Since its opening in March 2005, the MIXC has served some 20 million visitors, including large numbers of business-people and tourists from nearby Hong Kong. Phase II will include a 614,000-square-foot hotel, 258,000 square feet of additional retail, and 1.2 million square feet of residential.

Top left: Development's entry plaza and office building at key intersection.

Top middle: Customer connections through project's skylighted galleria.

Top right: Bird's-eye view of City Crossing.

Above: Entry plaza at dusk.

Bottom left: Gardened open spaces link street level to lower-level retail and cafes.

Photography: Tim Griffith.

Sasaki Associates Inc.

64 Pleasant Street
Watertown, MA 02472
617.926.3300
617.924.2748 (Fax)
www.sasaki.com
www.sasakigreen.com

77 Geary Street
San Francisco, CA 94108
415.776.7272
415.202.8970 (Fax)

Sasaki Associates Inc.

Dart CBD Transit Mall
Dallas, Texas

In 1988, Sasaki Associates was commissioned to design the Transitway Mall for Dallas's central business district, a 1.2-mile path that would accommodate a light rail system for DART (Dallas Area Rapid Transit). Facing a public that was skeptical about mass transit, DART officials took the unusual step of having an urban design/landscape/architecture firm, rather than an engineer, lead the design process. Working with the client and many user groups, Sasaki conceived the mall as a "Ramblas of the Southwest," referring to Barcelona's famous linear park. The 1.2-mile corridor includes four stations and is unified by street trees, distinctive streetscape treatments, and a public arts program. The light rail lines extend many miles into the outer city and suburbs. Since its completion in 1996, the Mall has grown from accommodating 1.4 million passengers initially to 17.5 million in 2005. Its landscape accoutrements are now firmly part of the city's character. And it is providing a real transportation alternative to those in the Dallas region looking for a more sustainable way of living.

Above: Light rail train at St. Paul stop.

Right: Aerial view.

Opposite top left: Evening at St. Paul station.

Opposite top right: Map of DART routes.

Opposite: Akard stop along Transitway Mall.

Photography: Craig Kuhner, Greg Hursley (right).

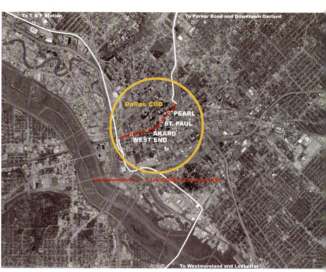

Sasaki Associates Inc.

Providence 2020
Providence, Rhode Island

Top: Plan of proposed transit spine, existing and proposed trolley lines, with 5-minute-walk circles around transit stations.

Right: Potential 2020 view of downtown core looking north.

Opposite: Providence 2020 Vision Plan.

Providence presents sharp contrasts – part gritty industrial city, part patrician state capital. Its numerous architectural landmarks include the capitol building, by McKim Mead & White, the Providence Athenaeum, and the stately buildings of Brown University. As in its bigger northern neighbor, Boston, civic leaders here have had the opportunity to redefine their city by relocating a highway that bisected its center. With the interstate rerouted, the pieces are in place for a renaissance: compact size, a long and potentially dynamic waterfront, numerous cultural amenities, and former industrial areas with vast loft spaces appealing to young students, artists, and entrepreneurs. There will also be sites for new residential and office buildings with spectacular views. Providence 2020 stresses easy walkability, mass transit options, and reuses of brownfield sites. It identifies opportunities for mixed-use development while promoting the unique characteristics of each distinctive district. A transit concept will integrate these districts, so that every resident senses an ownership of downtown. While elements of a working waterfront will remain, parks and waterfront promenades will serve and complement all districts. The proposed transit spine will parallel the waterfront greenway, with key stations setting the stage for transit-oriented development (TOD). Public parking garages are sited close to highway arrival points and spaced about one half mile apart to put them within a five-minute walk of most destinations. While Providence is known for pedestrian-friendly streets that calm traffic, the redeveloping industrial districts flanking it will need the positive identity of major streets. Municipal officials have already made great strides in "re-branding" the city as a focus of arts, research, and other intellectual pursuits. Providence 2020 will facilitate its transformation over the coming decades.

Sasaki Associates Inc.

Schenley Plaza
Pittsburgh, Pennsylvania

The redesign of this park has created the key civic open space for the regenerating Oakland district, where the public can enjoy both everyday pleasures and special events. Originally conceived as the gateway to the adjoining Schenley Park, Pittsburgh's counterpart to New York's Central Park, the plaza was defined by a rectangle of London Plane trees framing the Schenley Memorial Fountain. From the start, the plaza favored wide drives over pedestrian space, and it eventually became essentially a traffic circle. In the 1950s, it was converted to a parking lot, surrounded by heavily trafficked streets. Working with traffic consultants, Sasaki redesigned the surrounding roads to pedestrian scale, with on-street parking, thus assuring walkable connections to neighboring academic, institutional, and residential areas. To the original orthogonal axis of the space, the design adds a second, diagonal one that recognizes the likeliest point of pedestrian access to the plaza and the park beyond. Arrayed along this axis, flanking an unencumbered central lawn, are a series of retail pavilions and a carousel. Public toilets and a maintenance building are located in the tree bosques. By rebalancing the pedestrian/vehicular equation, the design encourages casual interaction involving a wide swath of the public and furthers the image of Oakland as a place where cars are not necessarily required, thus advancing two important and often overlooked components of environmental sustainability, social and economic sustainability.

Top left: Redesigned plaza, with Schenley Park and its memorial fountain beyond.

Top right: Schenley Plaza before redesign.

Far left: Activity along plaza's new diagonal axis.

Middle left: View of central lawn, showing seating and planting beds in foreground.

Left: Characteristic plaza pavilions.

Photography: Craig Kuhner (except for top right aerial.)

Sasaki Associates Inc.

New Jersey Urban Parks Master Plan Competition
Trenton, New Jersey

Sasaki led a team that was one of five finalists in a competition for a parks master plan to reconnect New Jersey's capital city of Trenton to its Delaware River waterfront. The plan encompasses several historic landmarks, including the State House, the Trenton Battle Monument, and portions of the Delaware & Raritan Canal. It would restore the riverfront landscape, now socially and ecologically lifeless, by removing hard river edges in favor of meandering shorelines with original vegetation. The city's grid would be extended to the water in a series of platforms, terraces, and piers providing places for walking, boating, fishing, picnicking, and outdoor events, while allowing the public to observe the river's changing water levels. Structures in the river's flood zone would work with natural forces, not against them, by being built on pilings or floating platforms. Low-maintenance meadows would replace mowed lawns. Programming would include public education on the natural environment and community participation in planting and maintaining it. This active reconnection of people with their river would foster both social and ecological sustainability.

Top left: Park master plan, with public routes to river in red.

Left: Rendering of pier and riverfront.

Bottom left: Park area with river tributaries.

SmithGroup

1850 K Street, NW
Suite 250
Washington, DC 20006
202.842.2100
202.974.4500 (Fax)
www.smithgroup.com

Ann Arbor
Chicago
Detroit
Los Angeles
Madison
Minneapolis
Phoenix
Raleigh-Durham
San Francisco
Washington, DC

SmithGroup

Downtown Detroit YMCA
Detroit, Michigan

This new YMCA continues Detroit's effort to re-stitch its frayed urban fabric. The project not only energizes the membership, but makes this energy visible in the city through transparent facades, allowing passersby to experience the activity inside. The program for the 100,000-square-foot building includes three major components: fitness, child care, and arts and humanities. Among the fitness facilities are a warm water pool, a leisure pool, and a four-lane lap pool, a full basketball court, two racquetball courts, and a banked indoor running track, plus a Wellness Center with aerobic and exercise equipment. Child care includes a short-term

"child watch" area along with a fully licensed all-day facility. The art and humanities area includes a 225-seat black box theater, a family arts center, and classrooms for media, literary, and visual arts. Architecturally, the structure can be thought of as a series of "tubes," stacked in a constructivist manner. Vertical-seamed metal panels form the tube walls, with glass curtain wall at the tube ends. Core elements are clad in brick. The tube with the Wellness Center has an all-glass exterior and hovers on tall concrete columns above the entrance plaza. The pools are a half-level below grade, with street-level windows allowing views in. On the interior, a split-level layout with glass partitions permits views up and down to different activity spaces. The centrally located climbing wall extends through multiple floors. The running track above the basketball court is on a level with the tracks and station of the elevated Detroit People Mover, just outside the transparent wall.

Far left: Wellness Center projecting over and sheltering main entry.

Top right: Building's transparent walls linking its interior to the city's streets.

Left: Gym's glass walls merging it into cityscape.

Above: Climbing wall rising through various interior spaces.

Photography: Justin Maconochie Photography.

SmithGroup

Discovery Communications Headquarters
Silver Spring, Maryland

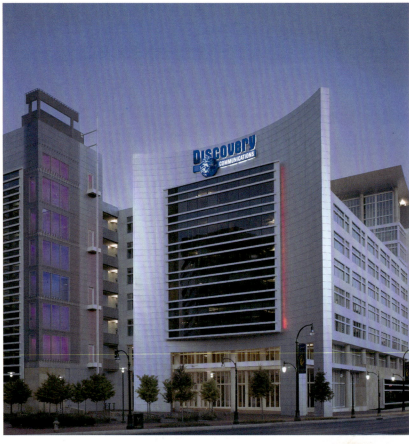

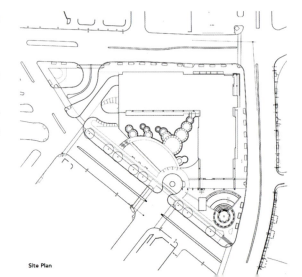

Top left: Stair tower, illuminated with changing colored light.

Top right: Discovery Plaza at main public entrance.

Right: Headquarters reestablishing street walls on two key streets.

Bottom right: Site plan.

Opposite: Atrium and employee lobby vividly visible at night.

Photography: Prakash Patel.

Occupying a previously empty 3.4-acre site at Silver Spring's major crossroads, this 580,000-square-foot headquarters complex initiates the rebirth of this old nearby suburb of Washington, DC. Some 2,000 employees occupy the building for this diversified media company, which provides international television programming, on-line services, and retail stores. The headquarters is designed to represent the company's cutting-edge work culture and its commitment to exploration, learning, and community involvement. It emphasizes the building's role as a gateway to Silver Spring and a center of community life. Configured as an L, the building has major facades on two thoroughfares, with a ceremonial entry at one end of the L. Exterior lighting creates a nighttime "show," using colors on the west façade and a continuous sequence through the spectrum on the sculptural stair tower. The site is also notable for its half-acre Discovery Plaza and one-acre Discovery Garden, with seating, informal play area, and interpretive gardens. These public spaces link the headquarters to the nearby Metrorail and Metrobus transit center while providing the community with an inviting space that contributes to its quality of life.

SmithGroup

Visteon Village Corporate Headquarters
Van Buren Township, Michigan

The headquarters for this automotive parts supplier takes the form of a village that embodies a variety of sustainable design strategies. Laid out on a 265-acre site, the complex includes administrative offices, development/delivery areas, technology/laboratory areas, employee and customer amenities, and support spaces totaling about one million square feet. The buildings are organized around three entry courts: Main Street, a piazza, and a promenade. About half of the office space has direct views of the on-site lake. The laboratory spaces and the remainder of the offices are along the far side of Main Street, forming intimate work neighborhoods for customer and employee teams. Vertical circulation is located in small towers that mitigate the perceived scale of the complex. An atrium lobby at the visitor entry functions as the village's living room. The lakefront piazza is the site of the dining hall, the customer presentation room, the auditorium and the village's focal campanile. One of the chief goals of the project was to improve the site's hydrology and habitat. A disused gravel quarry was transformed into a 37-acre lake which, because of its depth,

Top: Varied building forms seen across lake.

Above left: Lakeside buildings oriented for optimum solar access.

Above: Village plan around 37-acre lake.

Opposite: View showing distinctive details of buildings and terraces.

Photography: Justin Maconochie Photography; Laszlo Regos Photography (top).

SmithGroup

Left: Night view showing extent and configuration of glazing.

Bottom left: Street between buildings, with sidewalk amenities.

Photography: Laszlo Regos Photography.

could be used as a source of constant-temperature water for year-round space conditioning, as well as irrigation water when needed. On other parts of the property, remnant wooded wetlands were preserved, restored, and reconnected. The complex is designed to meet requirements for LEED certification. In addition to the site development strategies, the buildings feature high-performance daylighting and space conditioning. Along with exemplary orientation and highly efficient envelopes, the structures include raised-floor air delivery, providing for user comfort through stratification and individual controls. Water use and power demand are expected to be 35 percent and 30 percent below standard, respectively. Selection of construction and interior materials has focused on salvaged, recycled, rapidly renewable, and locally produced materials with a minimum of off-gassing.

Starck
Architecture + Planning

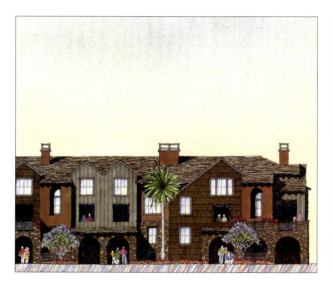

2045 Kettner Boulevard
Suite 100
San Diego, CA 92101
619.299.7070
619.295.8768 (Fax)
www.starckap.com

Starck
Architecture + Planning

Beacon Point at Liberty Station
San Diego, California

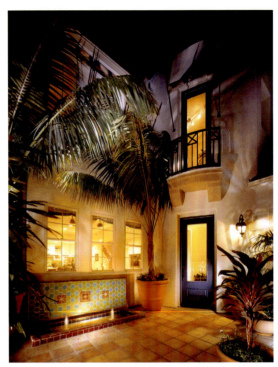

The location is the former Naval Station at Point Loma, now being redeveloped as a mixed-use waterfront community. For this 13.7-acre site, 140 units of detached single-family residences have been planned, ranging from 1,938 to 2,674 square feet, on lots 24 to 28 feet wide, with alley-accessed garages. The site plan respects the original layout of the site, as well as existing off-site street grids. A formally landscaped promenade running the length of the development integrates public access with community recreation.

Lush gardens, open spaces, seating areas, and bicycle paths reinforce the sense of a distinctive community. The architecture of the houses is inspired by the rich vocabulary of preserved Naval Station structures and the strong collection of 1920s and 1930s houses surrounding the site, which display Craftsman and Mediterranean styles. The new streetscapes are enhanced with architectural details that include balconies, wrought iron, tiled fountains and flower boxes, and striped awnings.

Opposite far left: Private entry court.

Opposite left: Aerial view of entire Beacon Point development, residential district outlined in yellow.

Right: House with Spanish-inspired front porch.

Below right: Corner house with evocative details.

Bottom: Row of characteristic new houses.

Photography: Lance Gordon, Lance Gordon Photography.

Site plan drawing: The City of San Diego Redevelopment Agency and The Corky McMillin Companies.

Starck
Architecture + Planning

Portico
San Diego, California

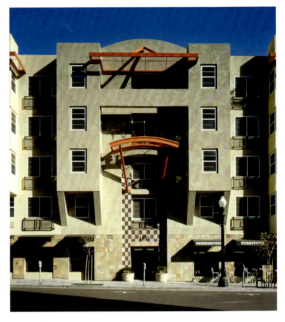

Located in the historic Little Italy section of downtown San Diego, this mixed-use building contains condominium units above underground parking and street-level retail. Varying in size from 432 to 999 square feet, the 84 condominium units are targeted to low- and mid-priced first-time homebuyers. Restrictions of city and community code and construction budget were accepted as aesthetic opportunities. Facades are broken up into vertical elements at neighborhood scale and enlivened by color changes, bright sunshades, and an identifying light tower above the street corner. Slate tiles enhance the façades at pedestrian level. Wide sidewalks allow retail to spill out onto outdoor eating areas. Landscaped open spaces on the interior of the block provide secluded areas for socializing and relaxing.

Top left: Overall view from street intersection.

Top right: Main lobby entry and storefronts.

Bottom left: Corner bay with lighted pinnacle.

Bottom right: Portion of façade with storefronts.

Photography: Owen McGoldrick, Owen McGoldrick Photography.

Starck
Architecture + Planning

Bridges at Escala,
San Diego, California

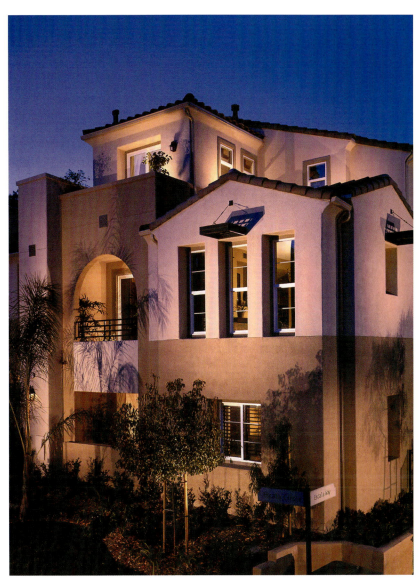

Vehicles can overwhelm higher-density residential projects. The design intent here was to contain the vehicles and focus attention on community features such as front doors and porches. The 136 units on this 7.6-acre site are laid out around motor courts that provide access to private two-car garages. The public street scene includes landscaped parkways and sidewalks. The three-story townhouses, varying from 1,186 to 1,705 square feet, include entries and some bedrooms on the first floor, main living spaces on the second, bedrooms and some decks on the third. Master bedrooms occupy the bridges at the ends of the buildings. Exteriors of the units are compositions of two- and three-story volumes, enlivened by arches, balconies, and trellises, with two plaster colors to enhance the play of forms.

Top left: End of typical building, with arched entry to motor court.

Top right: Characteristic townhouse front along landscaped walk.

Right: Site plan.

Photography: Jeffrey Aron, Aron Photography.

Site plan: Landscape architects Gillespie Moody Patterson, Inc.

Starck
Architecture + Planning

Ponto Beachfront Village
Carlsbad, California

Divided into Lots 1 and 2, the site is located at the key intersection of Carlsbad Boulevard and Avenida Encinas, directly across from Ponto Beach and Campground. The entire Beachfront Village is designed as a pleasant walking and bicycling environment, with extensive pedestrian plazas and paseos.

The mixed-use center on Lot 2 is designed as the lively focus of the development, with one- and two-story specialty shops, services, offices, and housing, along with restaurants appealing to both residents and visitors. A community facility and a wetland interpretive park add to the vitality of the center.

Lot 1 is developed with 126 townhouses, ranging in size from 1,125 to 1,501 square feet, on a gross area of 6.43 acres. Each three-story townhouse has an entry and two-car garage on the first floor, living areas on the second, and bedrooms on the third. The end units bridge over the motor court driveway. Architectural features and details reflect California coastal traditions. City zoning restrictions made it very difficult to attaining a residential density of 20 units per acre. They require open space of 200 square feet per unit, with a minimum dimension of 50 feet, and motor courts had to allow 34 feet

Below: Garage elevation, showing effect of grade change, with portion of mixed-use complex.

Bottom: Elevation of townhouse blocks, showing bridge over motor court entrance.

Opposite: Site plan of entire development.

Starck
Architecture + Planning

garage door to garage door, rather than a typical 24 to 28 feet. The mixed-use center on Lot 1 includes 10,700 square feet of restaurant, 21,200 square feet of retail, and 4,000 square feet of office space in seven live/work units. To accommodate this program in the available area, a four-level parking garage was required. With one level underground, the design exploits a 20-foot rise in grade to conceal most of the garage's bulk on two of its elevations, while adjoining mixed-use buildings abut it on the other two sides.

Top and above left: Elevations of mixed-use structures.

Left: Characteristic townhouse elevation.

Studio 39
Landscape Architecture, PC

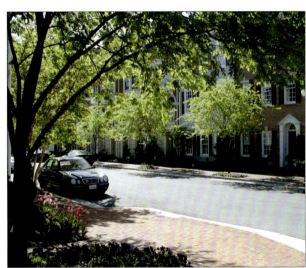

6416 Grovedale Drive
Suite 100-A
Alexandria, VA 22310
703.719.6500
703.719.6503 (Fax)
www.studio39.com

Studio 39
Landscape Architecture, PC

Workhouse Art Center at Lorton
Lorton, Virginia

Right: Proposed sculpture garden.

Below and Bottom right: Details of bioswale and rain garden.

Opposite top: Aerial view of central buildings and green.

Opposite bottom: Plan of entire property.

Drawings: Lisa Fraser (watercolor illustrations), Caitlin Dalton (rendered plan).

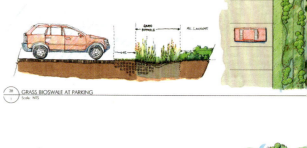

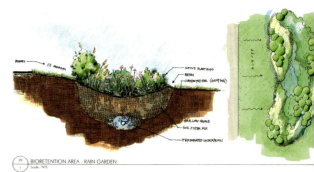

The planned conversion of the former Lorton Prison Center into a non-profit arts center starts with some impressive assets: a classically axial array of brick-walled, pitched-roof buildings surrounded by a scattering of outbuildings on casually planned terrain once used as a prison farm. The distribution of 215,703 square feet of buildings on 56 acres leaves 42 acres for the central green, outdoor studios, art gardens, playfields, and other public outdoor spaces. The adaptive reuse of the buildings themselves will include art studios, a restaurant district, and a residential colony. The zoning of the tract was converted from "residential-commercial" to "planned development commercial," and designs had to be approved by the local Architectural Review Board and the State Historical Architecture Department. The principal challenge is that public use of the property requires greatly expanded areas of parking. The visual impact of this parking is addressed with groves of shade trees, creatively organized to recall the orchards of prison farm years, yet placed to allow unobstructed views of the buildings from surrounding roads. Environmentally, the parking areas are mitigated by an L.I.D. (Low Impact Design) approach to landscaping, using selective placement of porous pavement, underground roof water collection facilities, storm water infiltration trenches, and rain gardens. Sidewalks are laid out to facilitate pedestrian movement through the complex and placed between buildings and parking areas to create appealing streetscapes.

Studio 39
Landscape Architecture, PC

Aurora Condominiums
Silver Spring, Maryland

For an apartment building renovation, a courtyard has been developed between the wings of the 10-story structure to function as a place of leisure and a passage from the lobby to the garage. Built atop a below-grade garage area, the landscaping had to employ green-roof methods and impose only limited added weight. The lobby-garage path is on a diagonal and is accented along one side by a line of cobalt blue glass block. A grove of Giant Cane Bamboo is on a berm to one side of the walk and provides a delicate backdrop. A line of bamboo extends in a gentle arc, dividing the plaza into two seating alcoves. Brushed stainless steel furniture enhances the minimalist modern design approach. Lighting required for security at night is provided by up-lights along the path, which emphasize the forms of the surrounding bamboo without disturbing adjoining apartments.

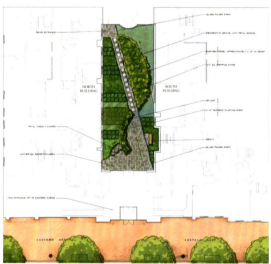

Top: Perspective of courtyard.

Far left: Detail of completed space, showing line of blue glass block.

Left: Partial plan of building, showing courtyard layout.

Photography: Jack Story.
Illustrations: Caitlin Dalton.

Studio 39
Landscape Architecture, PC

WRIT Rosslyn Center
Arlington, Virginia

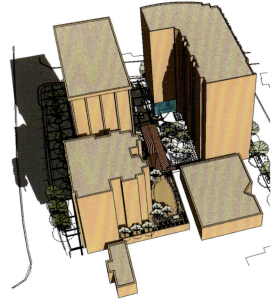

The landscaping on this 2.4-acre urban site is designed to integrate two new residential structures and an existing office building into one cohesive complex. Spaces between the buildings and atop underground parking structures are treated with landscape and hardscape, interwoven with three commissioned sculptures. The design features a 3,500-square-foot enclosed art atrium, a public plaza at the core of the site, an elevated semi-private courtyard, and streetscape treatment at the edges of the development. The developer, the landscape architect, and the building architect collaborated on the choice of commissioned art works, which are sited to maintain pedestrian interest as one moves into the complex. Tenant views of the landscape from above were carefully considered in the design of planting, pathways, and paving patterns.

Top left: View from the central enclosed atrium.

Left: Garden area with trellis.

Above: Bird's eye perspective, showing existing office building at upper left, two residential buildings, and low structures on adjoining sites.

Illustrations: Loren Helgason, Evan Timms.

Studio 39
Landscape Architecture, PC

Ford's Landing
Alexandria, Virginia

For an eight-acre planned development of townhouses, Studio 39 was involved from the initial planning phase in the location of open spaces, pedestrian linkages, site amenities, and sites for public art. The firm then proceeded with the design development, construction documentation, and construction administration for the entire project, including the tree-lined streets, brick-paved walks, three interior pocket parks, a tot lot, an entry plaza with a landscaped circle, and a boardwalk park along the Potomac. The approximately 2,000-foot of boardwalk features educational plaques and a sculptural shade structure. Since the property had been occupied by an old Ford Motor Company manufacturing plant, removal of contaminated soil was essential, and preservation of wetlands adjoining the site required dredging and

Above: Aerial view of waterfront portion of development.

Opposite top: Axial approach to boardwalk, with shade structure.

Opposite bottom: Site plan of entire project.

Photography: Jacob Kuntz.
Illustrations: Lisa Fraser.

Studio 39
Landscape Architecture, PC

bank reclamation. Sustainable strategies included recycling of concrete paving on the site as ballast below the boardwalk and landfill for new waterfront construction areas. The main goals of the site planning and design were to maximize public use of the riverfront – making the most of limited open area – and seamless connections to adjacent waterfront properties.

Top left: Characteristic streetscape.

Top right: Pocket park on interior of site.

Left: Corner park on interior of site

Photography: Jack Story, Jacob Kuntz.

2700 Promenade Two
1230 Peachtree Steet NE
Atlanta, GA 30309
404.888.6600
404.888.6700 (Fax)
www.tvsa.com

Atlanta
Chicago
Dubai
Shanghai

Thompson, Ventulett, Stainback & Associates (TVS)

Washington Convention Center
Washington, DC

Situated in the heart of Washington, this 2.1-million-square-foot complex – the city's largest building – preserves the L'Enfant master plan grid with its suggestion of three connected buildings, allowing two streets to pass through and maintain the continuity of the community. The design team collaborated continuously with civic groups and worked within tight urban constraints, which required placing one story below grade to meet the city's 110-foot height restriction. The building contains 700,000 square feet of exhibit halls, 150,000 square feet of meeting space, and 60,000 square feet of ballroom. The placement of concourses, registration areas, and lobbies with generous glass areas around the perimeter reveals the center's activity to the streets outside. The entry atrium façade, while monumental in scale and modernist in design, is complementary to the old Carnegie Library opposite it. The complex has reestablished the city's place in the convention and meeting world and created new economic opportunity for surrounding neighborhoods. It has received a prestigious Honor Award for Architecture from the American Institute of Architects and the Urban Land Institute's 2006 Award of Excellence.

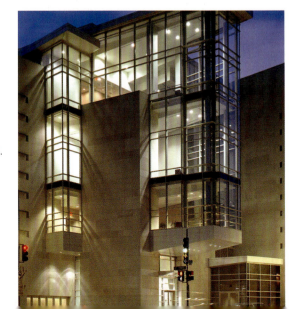

Top left: Entry façade at night, showing atrium as part of cityscape.
Top right: Convention Center with Carnegie Library in foreground.
Middle right: Circulation node with suspended sculpture by Donald Lipski.
Above right: Architectural massing to suggest series of connected buildings.
Right: Circulation spaces in crow's nests overlooking city.
Opposite: Entry atrium.
Photography: Brian Gassel/TVS.

Thompson, Ventulett, Stainback & Associates (TVS)

Technology Square, Georgia Institute of Technology
Atlanta, Georgia

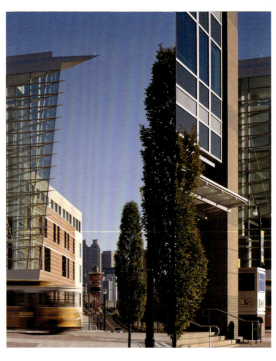

Technology Square reaches across the downtown connector from Georgia Tech into Atlanta's midtown business district, establishing a new gateway to the campus. Cantilevered lanterns signal to passersby on the interstate and those passing through the complex that something new and special has arrived within Midtown Atlanta. Replacing three city blocks of parking lots and overgrown vegetation, the project includes educational and retail facilities and a hotel/conference center. Street-level shops appealing to students and the public bring new life to the streetscape. The College of Management, one of the development's five buildings, is the second LEED Silver Level building in the state of Georgia and the 14th in the nation. It exceeded LEED requirements for both recycled content and locally produced materials. Other environmentally conscious features include effective use of daylight for interiors, reduction of water use, optimized energy performance, construction waste management, and alternative transportation. Technology Square received the Urban Land Institute (ULI) 2004 Award of Excellence.

Above left: Cantilevered projection that identifies complex from distance.

Above right: Street scene with retail geared to students and public.

Top right: Detail of projecting glazed volume.

Right: Lantern-like forms marking gateway to complex.

Opposite top left: Characteristic mix of glazed volumes and masonry-clad blocks.

Opposite top right: Complex mediating between campus and business district.

Photography: Brian Gassel/TVS.

Thompson, Ventulett, Stainback & Associates (TVS)

Plaza Norte
Santiago, Chile

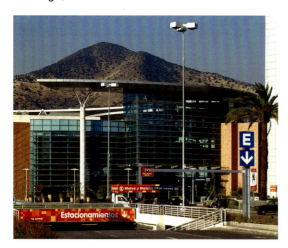

Far left: Plaza Norte's dramatic entrance plaza.

Left: Retail, cultural, and entertainment venues around central garden at Plaza Norte.

Photography: Brian Gassel/TVS.

Not just a retail mall, Plaza Norte incorporates medical office towers, big-box retailers, cultural and performing arts venues, and entertainment, along with the traditional fashion, food, and specialty stores. Within its total of 700,000 square feet it includes 75,000 square feet of medical offices and 75,000 square feet of other office space. A large central plaza/garden provides a sense of place for the center, a gathering place, and access to Santiago's subway system.

Short Pump Town Center
Richmond, Virginia

Drawing on Richmond's rich historical context and community spirit, TVS designed the nation's first major two-story, open-air shopping center. The project's 1.4 million square feet include an upscale mix of anchor stores, specialty shops, and restaurants. The pedestrian experience is focused on two main streets, which converge on a town square with grassy lawns for picnics, sidewalks for strolling, a gazebo for local events, an interactive fountain, and a playground. Way-finding is supported by distinctive graphics and unique iconography, which includes silhouetted sculptural busts of the eight U.S. presidents from Virginia.

Above left: Short Pump's town square with lawns, fountains, and other amenities.

Above: Two-story open galleries around courtyard at Short Pump Town Center.

Photography: Brian Gassel/TVS.

Thompson, Ventulett, Stainback & Associates (TVS)

Triangle Town Center
Raleigh, North Carolina

The natural beauty and cultural background of the state are recalled in the approach to the center via a tree-lined boulevard with a stream running through it. The boulevard feeling continues inside in the central promenade under an arched glass roof. The stream becomes an 18-foot waterfall at the center court, then re-emerges outside the 30-foot-high glass walls of the food court, creating a creek-side dining and entertainment plaza. The 1.2-million-square-foot complex is laid out as five different "buildings," each situated around a central "street."

Left: Boulevard-like interior arcade at Triangle Town Center.

Above: Entering Triangle Town Center along tree-lined stream.

Below left: Triangle Town Center's creek-side dining terrace.

Photography: Brian Gassel/TVS.

Thompson, Ventulett, Stainback & Associates (TVS)

Georgia Aquarium
Atlanta, Georgia

The design of the aquarium purposely departs from Atlanta's street grid and its characteristic architectural aesthetic. The structure's expressive architectonic forms swirl about a central blue volume, conceived as a vessel of conservation, preservation, and education. The complex represents, metaphorically, a large body of water undulating against the street that borders Centennial Park. Given Atlanta's landlocked location, the design team focused on a sense of immersion for the interiors, enveloping visitors in an underwater world and bringing them face-to-face with incredible ocean creatures. The aquarium presents 60 exhibits, including over 100,000 specimens of over 500 different species. With tanks totaling 8 million gallons – 6.2 million gallons in the largest one – the facility nevertheless uses less water than a typical office building.

Top right: Aquarium seen from Centennial Park.

Above left: Central atrium.

Above: Entrance canopy and central volume.

Left: Aerial view of aquarium.

Bottom left: Beluga shale exhibit.

Photography: Brian Gassel/TVS.

Torti Gallas and Partners

1300 Spring Street
4th Floor
Silver Spring, MD 20910
301.588.4800
301.650.2255 (Fax)
www.tortigallas.com

523 West 6th Street
Suite 212
Los Angeles, CA 90014
213.607.0070
213.607.0077 (Fax)

Torti Gallas and Partners   Sustainable Urbanism

Concentrating development in existing urban centers is not only a linchpin of the sustainability movement, but extends the life of our cities, among our most important resources. Infill development throughout Washington, DC is attracting new residents and rejuvenating neighborhoods.

The Ellington, Washington, DC
Located along U Street, Washington's former "Black Broadway," this high-rise apartment building includes 15,000 square feet of new ground-floor retail and 186 dwelling units, including apartments and luxury lofts. The vertical sign at its corner is reminiscent of the old theater signs that were once signature elements of this street.

3030 Clarendon Boulevard, Arlington, Virginia
This two-block, mixed-use infill development at the Clarendon Metro stop will be a LEED-rated development. Included are 54,000 square feet of retail, 83,000 square feet of offices, and 244 residential units. The massing of the building will create a landmark feature at the corner, with a distinctive curved façade of brick and glass.

City Vista, Washington, DC
Adjacent to the new Convention Center, City Vista is a major building block in the city's plans to attract new residents to the quarter. The high-rise building will contain 632 housing units, 20 percent of them affordable, as well as a Safeway grocery store and other neighborhood retail.

Top left: The Ellington.
Top right: 3030 Clarendon Boulevard.
Middle and bottom right: City Vista.
Photography: Steve Hall @ Hedrich Blessing.
Renderings: Interface Multimedia.

Torti Gallas and Partners

Columbia Heights, Washington, DC

The redevelopment of Columbia Heights, one of Washington's oldest neighborhoods, has restored the community to its former vibrancy. Three projects, clustered around the existing Metro station, will animate 14th Street with new retail and residential activity.

Highland Park
This new high-rise building contains 229 apartments and 19,000 square feet of ground-floor retail. Inspired by the local Art Deco style, the structure forms a new piazza at the subway entrance and incorporates an existing Metro cooling tower in its façade.

Kenyon Square
A 75-unit affordable senior housing complex will be included in this high-rise building, along with 153 condominiums. Ground-floor retail will include cafes with outdoor seating. The façade features a mix of styles designed to animate the long street front.

Park Triangle
Adjacent to the historic Tivoli Theater, this new mid-rise building contains 18,000 square feet of retail and 131 apartments, including efficiencies and lofts. The building's bays, cornices, and façade composition are inspired by the architecture of the adjacent theater and Riggs Bank.

Park Place Condominiums, Washington, DC

Located at the Georgia Avenue Metro station, this new building contains 148 apartments over 17,000 square feet of retail, with seven additional fee-simple townhouse units on the residential street around the corner. The long avenue façade is designed in an eclectic set of languages, featuring an expressive Modernist curve at the Metro end of the complex.

Top right: Columbia Heights site plan showing all three projects.
Top left: Highland Park.

Left column, top to bottom: Kenyon Square, Park Triangle, Park Place Condominiums.

Renderings: Interface Multimedia.

Torti Gallas and Partners

TODs and Town Centers

Compact, pedestrian-friendly mixed-use transit-oriented developments — TODs – are essential to any city and fundamental elements of the Smart Growth and Sustainability movements.

Twinbrook Commons, Rockville, Maryland
A vibrant new mixed-use neighborhood around an existing Metro station will include 1,295 residential units, 620,000 square feet of offices, and 160,000 square feet of retail. A variety of high- and low-rise buildings will be laid out around a new set of streets and a public piazza. The development will be seamlessly integrated with the existing neighborhood.

Shirlington Village, Arlington, Virginia
A multi-block development in the heart of Shirlington, Arlington's arts and entertainment district, will transform a suburban strip center into a mixed-use village including 241 residential units, a Harris Teeter grocery, and other retail.

The Residences at the Greene, Beaverton, Ohio
This landmark building is the centerpiece of The Greene, an 800,000-square-foot mixed-use core for Greene County. A precedent-setting project, The Residences introduces a mix of luxury residential units at the heart of a new lifestyle center.

Cooper's Crossing, Camden, New Jersey
A master plan for the Camden waterfront will transform it into a vibrant mixed-use district including 1,602 residential units, 398,000 square feet of offices, 78,000 square feet of flex space, and 56,250 square feet of retail. Currently vacant parcels will be redeveloped into entertainment venues to complement existing facilities, creating the critical mass of activity that will give this riverfront a bold new image.

Top and upper middle right: Shirlington Village.

Right and lower middle right: Twinbrook Commons.

Below: The Residences at the Greene.

Photography: Steve Hall @ Hedrich Blessing.

Renderings: Interface Multimedia.

Left and below left: Cooper's Crossing.

Below: Plan and two perspectives, Lansdowne Village

Renderings: © Michael B. Morrissey MRAIC.

Lansdowne Village Greens, Loudoun County, Virginia

A new town center will combine a variety of residential units, totaling 880, with 150,000 square feet of retail, 80,000 square feet of offices, and 55,000 square feet of flex space. Anchored by a mixed-use main street and a new town square, this compact project will transform a suburban strip mall into a real village, establishing a new development paradigm for Loudoun County.

Torti Gallas and Partners

Sustainable Neighborhoods

Compact, mixed-use, pedestrian-friendly neighborhoods with interconnected streets and open spaces are core components of sustainability. Equally important are environmentally sensitive design and the inclusion of a range of residential units to accommodate the economic and social diversity that sustains communities over time.

Military Family Housing, Fort Belvoir, Virginia

Building on the history of this long-established post, this new military family neighborhood combines rich urban and architectural traditions with the best aspects of traditional American towns. A new mixed-use Main Street, the first of its kind on a military installation, creates a vital amenity at the heart of the post. Included are 1,630 new residential units, including live/work units in the town center, and the renovation of 178 historic units.

Martin Luther King Plaza, Philadelphia, Pennsylvania

This development replaces a former public housing project with 349 new units in a mixed-income, mixed-use neighborhood. Substantial infill in the surrounding Hawthorne district, combined with on-site development, will create the stability needed to sustain the area over time.

Centergate Baldwin Park, Orlando, Florida

Variegated rowhouses, courtyard apartments, and single-family units – a total of 314 units in four distinct architectural styles – are combined to create a rich diversity. Flex buildings, with ground-floor units that can be converted to office or retail uses, establish a mixed-use Main Street.

Salishan, Tacoma, Washington

Located on a sensitive watershed connected to precious salmon spawning grounds, this development marries traditional neighborhood principles with sustainable environmental practices. Its 1,180 residential units will include 704 rental, 356 for sale, 105 senior, and 15

Top three photos: Fort Belvoir Military Family Housing.
Right: Martin Luther King Plaza.
Photography: Steve Hall @ Hedrich Blessing; Torti Gallas (upper right).

Habitat for Humanity houses, all with a regionally-inspired Craftsman design theme. A mixed-use program transforms this former public housing site into a neighborhood sustainable in all its aspects – social, economic, environmental, and cultural.

Left and below left: Centergate Baldwin Park.
Photos below: Salishan.
Photography: Steve Hall @ Hedrich Blessing

Torti Gallas and Partners

Sustainable Urbanism in Los Angeles

Left, top to bottom: Perspective, Vantaggio of Baldwin Hills; before and after images, Downtown Upland; aerial perspective and representative elevation, Taylor Yards.

While the themes of sustainability and urbanism are universal, their adaptation to local cultures and traditions is essential to their success. The Los Angeles office of Torti Gallas has harnessed the firm's considerable experience in these arenas, focusing it on the architectural and urban issues and rich culture of this city and region.

Vantaggio of Baldwin Hills, Los Angeles, California

This 18-unit project in the city's southwestern hills will transform a vacant lot into an exciting residential building capitalizing on spectacular views of downtown. The new structure will fit seamlessly into the existing neighborhood and is within a height envelope prescribed by city ordinance. The design of the structure, which will be LEED certified, is based on the hip yet elegant design image of Los Angeles.

Downtown Upland Master Plan, Upland, California

This master plan will revitalize downtown Upland, a historic Inland Empire community, transforming it into a vital, pedestrian-friendly town center. A "park once" strategy, combined with mixed-use infill development of vacant parcels, leverages the Metro-Link Commuter Rail station and will re-animate existing streetscapes, forming a cohesive and interconnected public realm.

Taylor Yards Redevelopment, Los Angeles, California

On a highly prized site along the Los Angeles River, a new master plan will create a vibrant transit-oriented mixed-use development. The program includes 25,000 square feet of ground-level retail with approximately 450 residential units – both market-rate and affordable. An intricate layout of streets connects to a central linear green, which links the new retail street to an adjacent park. In transforming the river edge, the project offers a paradigm for future development along its banks.

Project Credits

Exceptional care has been taken to gather information from firms represented in this book and transcribe it accurately. The publisher assumes no liability for errors or omissions in the credits listed below.

ANKROM MOISAN ARCHITECTS

Chown Pella Lofts
Client: Carroll Aspen Investments
Principal consultants:
Ankrom Moisan Architects, architectural design, interior design
Design-Build, Jacobs Heating, mechanical consultant
Design-Build, Oregon Electric, electrical consultant
KPFF Consulting Engineers, structural consultant
KPFF Consulting Engineers, civil consultant
Airo Design, landscape architect

Bullseye Glass
Client: Bullseye Glass
Ankrom Moisan Architects, architectural design, interior design
AAI Engineers, structural consultant

McKenzie Lofts
Client: Carroll Investments
Principal consultants:
Ankrom Moisan Architects, architectural design
Design-Build, Jacobs Heating, mechanical consultant
Design-Build, Oregon Electric, electrical consultant
KPFF Consulting Engineers, structural/civil consultant

Riverstone Condos
Client: Hoyt Street Properties
Principal consultants:
Ankrom Moisan Architects, architectural design, interior design
Interface, mechanical and electrical consultant
Kopcynski & Associates, structural consultant
David Evans & Associates, civil consultant
Perron Collaborative, landscape architect

Tanner Place
Client: Hoyt Street Properties
Principal consultants:
Ankrom Moisan Architects, architectural design, interior design
Interface Engineering, mechanical consultant
Interface Engineering, electrical consultant
Kramer Gehlen Associates, structural consultant
Landscape Architect: Nevue Ngan & Associates, landscape architect

The Gregory
Client: Carroll Investments
Principal consultants:
Ankrom Moisan Architects, architectural design, interior design
Interface Engineering, mechanical consultant
Interface Engineering, electrical consultant
Kramer Gehlen Associates, structural consultant
W&H Pacific, landscape architect

Lovejoy Station
Client: Housing Authority of Portland and Onder Development
Principal consultants:
Ankrom Moisan Architects, architectural design, interior design
R & W Engineering, bidder design, mechanical/electrical
Harper Houf Petersen and Righellis, civil engineer
Associated Consultants, Inc., structural consultant
Simp.L PC, landscape architect

Marshall Wells Lofts
Client: Evergreen Northern
Principal consultants:
Ankrom Moisan Architects, architectural design, interior design
Interface Engineering, mechanical and electrical consultant
Harper Houf Petersen and Righellis, civil engineer
KPFF Consulting Engineers, structural consultant
Perron Collaborative, landscape

Pearl Clinic
Client: Pearl Clinic, tenant improvement
Principal consultants:
Ankrom Moisan Architects, architectural design, interior design, sustainable design
Bidder design, mechanical consultant
Bidder design, electrical consultant

Bridgeport Condominiums
Client: Hoyt Street Properties
Principal consultants:
Ankrom Moisan Architects, architectural design, interior design
Interface Engineering, mechanical consultant
Oregon Electric Group, electrical consultant
Kramer Gehlen & Associates, structural consultant

David Evans & Associated, civil consultant
Steve Koch Landscape, ASLA, landcape architect

Elizabeth Lofts
Client: Carroll-Aspen Elizabeth, LLC
Principal consultants:
Ankrom Moisan Architects, architectural design, interior design, sustainable design
Interface Engineering, mechanical and electrical consultant
Kramer Gehlen & Associates, structural consultant
Civil and Landscape Consultantt: W&H Pacific, civil and landscape consultant

10th and Hoyt Apartments
Client: Trammell Crow Residential
Principal consultants:
Ankrom Moisan Architects, architectural design, interior design, sustaintable design
Interface Engineering, mechanical and electrical consultant
Kramer Gehlen & Associates, structural consultant
Steve Koch, ASLA, landscape architect

Burlington Apartments
Client: Pearl Block LLC
Principal consultants:
Ankrom Moisan Architects, architectural design
McKinstry Co., design/build mechanical
New Tech Electric, Inc., design/build electrical
Kramer Gehlen & Associates, structural consultant
Steve Koch, ASLA, landscape architect

The Pinnacle Condominium
Client: Hoyt Street Properties
Principal consultants:
Ankrom Moisan Architects, architectural design
Interface Engineering, mechanical and electrical consultant
Kramer Gehlen & Associates, structural consultant
Steve Koch, ASLA, landscape architect

Sitka Apartments
Client: Praxis Partners
Principal consultants:
Ankrom Moisan Architects, architectural design, sustainable design
Interface Engineering, mechanical and electrical consultant
Kramer Gehlen & Associates, structural consultant
Steve Koch, ASLA, landscape architect

937 Condominiums
Client: 937 Group, LLC, an affiliate of W&K Develpment, LLC
Principal consultants:
Ankrom Moisan Architects, architect of record
Holst Architecture, design architect
Hunter Davisson, design/build mechanical
Peninsula Plumbing & New Tech Electric, design/build electrical
Kramer Gehlen & Associates, structural consultant
Steve Koch, ASLA, landscape architect

Mirabella Seattle
Client: Pacific Retirement Services
Principal consultants:
Ankrom Moisan Architects, architectural design
McKinstry, mechanical consultant
SASCO, electrical consultant/low voltage systems:
Cary Kopczyynski & Company, structural consultant
Robert H. Foster Consultants, landcape architect

Rollin Street Flats
Client: City Investors III
Principal consultants:
Ankrom Moisan Architects, architectural design
Holaday-Parks, mechanical/plumbing
Gerber Engineering and Prime Electric, electrical consultant
Cary Kopczynski & Co., Inc. P.S., structural consultant
Brumbaugh & Associates, landscape architect

One South Market
Client: Spring Capital Group
Principal consultants:
Ankrom Moisan Architects, architectural design
Interface Engineering, mechanical and electrical consultant
Magnusson Klemencic Associated, structural consultant
Kla-Koch Landscape Architecture, landcape architect

Arterra
Client: Capstone Partners LLC
Principal consultants: Ankrom Moisan Architects, architectural design, sustainable design
Interface Engineering, mechanical and electrical consultant
Kramer Gehlen & Associates, structural consultant
Lango Hansen, landscape architect

Luma
Client: The South Group
Principal consultants: Ankrom Moisan Architects, architectural design, sustainable design
Thermalair, mechanical designer
Morrow Meadows Corp., electrical designer

Magnusson Klemencic Associates. structural consultant
ah'bé landscape architects, landscape architect

Elleven
Client: The South Group
Principal consultants:
Ankrom Moisan Architects, architectural design, sustainable design
Glumac, design build/mechanical and electrical
Nabih Youssef and Associates, structural consultant
ah'bé Landscape Architects, landscape architect

L Street Lofts
Client: SKK Development
Principal consultants:
Ankrom Moisan Architects, architectural design
Interface Engineering, mechanical and electrical consultant
Kramer Gehlen and Associates, Inc., structural consultant
Land Architecture, landscape architect

ANNEX|5

Lake Bluff Tower
Client: Mandel Group, Inc.
Principal consultants:
annex|5, master planning, architecture
A. Epstein and Sons International, Inc., engineering

China Science and Technology Museum
Client: Peoples Republic of China
Principal consultants:
annex|5, master planning, architecture
Tsinghua Institute, architect and engineer of record

Beijing Century City East
Client: Beijing Century Town Estate Development Co., LTD.
Principal consultants:
annex|5, architectural design
ARUP, structural design consultants

Beijing CBD Office Park
Client: Beijing Longzeyuan Real Estate Company
Principal consultants:
annex|5, architectural design consultant
Beijing Institute of Architecture and Design (BIAD), architect and engineer of record
LA International, design architect

ARCHITECTS ORANGE

Riverside Plaza
Client: Riverside Plaza, LLC/Litchfield Advisors

Principal consultants:
Architects Orange, architectural design
Graphic Solutions, signage and graphic design
Konsortium 1, electrical/lighting design
RHA Landscape Architects, landscape architecture
KCT Consultants, Inc., civil engineering

Piemonte at Ontario Center
Client: Panattoni Development
Principal consultants:
Architects Orange, master plan and architectural design
EPT Landscape Design, landscape
SB & O, civil engineering

Seattle Premium Outlet Center
Client: Chelsea Property Group, a Simon Company
Principal consultants:
Architects Orange, architectural design
Stearns Architecture, conceptual design, materials selection, detailing
Ken Large, Landscape Architect, landscape
Fabritec Structures, design development, design and fabrication of all tensile fabric structures

Alexan Pacific Grove
Client: Trammell Crow Residential
Principal consultants:
Architects Orange, architectural planning and design
MJS Landscape Design Group, landscape
Ruzika Lighting, lighting

BALDAUF CATTON VON ECKARTSBERG ARCHITECTS

The Ferry Building Marketplace
Client: The Port of San Francisco, owner; Equity Office Properties Trust, Wilson Meany Sullivan LLP, Primus Infrastructures LLP, developers
Principal consultants:
SMWM, architect
Baldauf Catton von Eckartsberg Architects, retail architect
Page & Turnbull, preservation architects
Tom Eliot Fisch, architect for the Port Commission Hearing Room
Horton Lees Brogden, lighting design
Arias Associates, building signage design
Plant Construction Company, LP, general contractor

Oxbow Public Market
Client: Oxbow Management
Principal consultants:
Baldauf Catton von Eckartsberg Architects, architect
Kellie Carlin, landscape design
Revolver Design, lighting design

MBA, structural engineer
Riechers, Spence and Associates, civil engineer

Mercato
Client: Trono Group, owner/developer
Principal consultants:
Baldauf Catton von Eckartsberg Architects, design architect
BBT Architects, architects of record
iREX, landscape architect
BE, civil engineer

Treasure Island Master Plan 2005
Client: Treasure Island Community Development, LLC, Lennar, Kenwood Investments, and Wilson Meany Sullivan, developers
Principal consultants:
Skidmore, Owings & Merrill, architects and urban design
SMWM, planning, urban design, and community development
Baldauf Catton von Eckartsberg Architects, all aspects of retail design and interre lated open spaces
Conger Moss Guillard with Thomas Leader Studios, landscape architect
Hornberger + Worstell, hotel architecture
Arup, sustainable design and transportation
Korve Engineering, civil engineer
Treadwell & Rollo, Engeo, geotechnical design
Concept Marine Associates, aquatic transit

BEELER GUEST OWENS ARCHITECTS

Eastside
Client: Post Properties, Inc.
Principal consultants:
Beeler Guest Owens Architects, architect
Perry Hescock & Associates, mep engineer
McHale Engineering, Inc., structural engineer
Graham Associates, Inc., civil engineer
Bosse & Turner, landscape design

3949 Lindell Boulevard
Client: HSAD
Principal consultants:
Beeler Guest Owens Architects, architect
KAI Design and Build, mep engineer and structural engineer
J. R. Grimes, civil engineer
Lewisites, Inc., landscape designer

Mercer Square
Client: Post Properties, Inc.
Principal consultants:
Beeler Guest Owens Architects, architect
Perry Hescock & Associates, mep engineer

Sterling Engineering & Design Group, structural engineer
Bury + Partners, civil engineer
Land Design Partners, landscape designer

Artessa at Quarry Village
Client: Embrey Partners, Ltd.
Principal consultants:
Beeler Guest Owens Architects, architect
Basharkhah Engineering, mep engineer
Sterling Engineering & Design Group, structural engineer
Pape Dawson, civil engineer
Enviro Design, landscape designer

The Triangle
Client: Cencor Urban
Principal consultants:
Beeler Guest Owens Architects, architect
Perry Hescock & Associates, mep engineer
Hunt & Associates, Inc., structural engineer
Bury + Partners, civil engineer
Coleman & Associates, landscape designer

BOOTH HANSEN

Cook County Circuit Courthouse
Client: Cook County
Principal consultants:
Booth Hansen, architect, architect of record (façade and public spaces, courtrooms, corridors, entry)
George Sollitt Construction Co, general contractor
WMA Consulting Engineers, Ltd., mep engineer
Wiss, Janey, Elstner Associates, Inc., structural engineering
McClier, civil engineer
Kroll, security consultants
Sako & Associates, Inc., audio/visual consultants
Wolff Clements and Associates, Ltd., landscape architect

MB Financial Bank Headquarters
Client: MB Financial Bank
Principal consultants:
Booth Hansen, site selection, feasibility, architectural design, interiors
Pepper Construction, general contractor
V3 Consultants, civil engineers
Cosentini Associates, mep/fp engineers
Thornton Tomasetti Engineers, structural engineers
Jacobs/Ryan Associates, landscape architect

Palmolive Building
Client: Draper and Kramer
Principal consultants:
Booth Hansen, feasibility, architectural design, architectural interiors

Walsh Construction, general contractor
Pepper Construction, general contractor
WMA Consulting Engineers, Ltd., mep engineers
Thornton Tomasetti Engineers, structural engineers

The Joffrey Tower
Client: Smithfield Developers
Principal consultants:
Booth Hansen, architectural design, interiors
Smithfield Construction Group, general contractor
Advance Mechanical Systems, mechanical/plumbing engineers
Innovative Building Concepts, electrical engineer
Thornton Tomasetti Engineers, structural engineer
MDJ Engineering, civil engineer
Wolff Landscape Architecture, landscape architect

BRAUN + YOSHIDA ARCHITECTS

Belle Creek
Client: New Town Builders
Principal Consultants:
B+Y Architects, site planning and architectural services
Mandil, exterior color design
Humphries Poli, architects of apartments and family center

Belmar
Client: McStain Neighborhoods
Principal Consultants:
B+Y Architects, architectural services
MNA, Inc., structural consultants
Mandil, exterior color design

The Brownstone Collection
Client: Shea Homes
Principal Consultants:
B+Y Architects, architectural and site planning services
RLA Design, master plan
The Lund Partnership, civil engineering
AKM Engineering Consultants, Inc., structural engineering
Sage Design Group LLC, landscape architecture
Hamilton Jones Design Group, sales office/model interior design
Mandil, exterior color design

The Architect Collection at Stapleton
Client: Harvard Communities
Principal consultants:
B+Y Architects, architectural services
Complete Engineering Services, Inc., structural engineer
Mandil, exterior color design

The Lofts at Stapleton
Client: The Fullerton Company
Principal consultants:
B+Y Architects, architectural services
MNA, Inc., structural engineer
Palace Construction, general contractor
Captivating Design Service, sales office/model interior design

The Townlofts at Stapleton
Client: The Fullerton Company
Principal consultants:
B+Y Architects, architectural services
Weintraub Organization Limited, structural engineer
C&D Management, general contractor

Roslyn Court at Stapleton
Client: Forest City, TP Development
Principal consultants:
B+Y Architects, architectural and site planning services
MNA, Inc., structural engineer
Norris Dullea, landscape architecture
Shaw Construction, general contractor

Tuck-Under Avenue Row Houses at Stapleton
Client: McStain Neighborhoods
Principal consultants:
B+Y Architects, architectural services
MNA, Inc., structural engineer
Mandil, exterior color design

CARTER & BURGESS

Towson Town Center
Client: General Growth Properties, Inc.
Principal consultants:
Carter & Burgess, Inc., architectural design, environmental graphics
Timothy Haahs & Associates, Inc., parking garage design and structural
Hope Furrer Associates, Inc., structural
E&S Construction Engineers, Inc., mechanical
B&R Construction Services, Inc., electrical
The Lighting Practice, Inc., lighting consultant
Mahan/Rykiel Associates, Inc. interior landscape architect
DMW, Inc., civil engineer and exterior landscape architect

Boca Raton Mixed-Use
Client: Confidential
Principal consultants:
Carter & Burgess, master planning, architectural design, civil engineering, survey, programming

Las Colinas Entertainment
Client: Texas Spirit, LLC
Principal consultants:
Carter & Burgess, master planning and architectural design

Ibn Battuta Mall Phase II
Client: Confidential
Principal consultants:
Carter & Burgess, master planning, architectural design, mep, structural

CMSS ARCHITECTS

Town Center of Virginia Beach
Client: Armada Hoffler Development; City of Virginia Beach
Principal consultants:
CMSS Architects, PC, master planning, architecture, landscape architecture, interior design
Armada Hoffler Construction Company, general contractor
Divaris Real Estate, broker
MSA, PC, civil engineer
Mathew J. Thompson, III, Consulting Engineers, Inc., mep engineer
McCallum Testing Laboratories, geotechnical

Kincora
Client: Norton Scott, AIG, Tritec (development group)
Principal consultants:
CMSS Architects, PC, urban planning

City Center at Oyster Point
Client: Newport News City Center, LLC (owner); HL Development Services Group, LLC (developer); Economic Development Authority of the City of Newport News (public client)
Principal consultants:
CMSS Architects, PC, master planning, architecture, landscape architecture, interior design
Mastin & Kufka, marketing

Rocketts Landing
Client: WVS Companies, Rocketts Landing LLC (owners/developers)
Principal consultants:
CMSS Architects, PC, master planning, architecture, landscape architecture, interior design
Wiley & Wilson, structural
Williams, Mullen and Dobbins, zoning, environmental, general counsel
Integra Realty Resources, real estate consultants

DERCK & EDSON ASSOCIATES

Lititz Watch Technicum
Client: HPR Lititz, L.P.
Principal consultants:
Derck & Edson Associates, site selection, feasibility including concept design, site design, storm water management design, municipal approvals, vehicular and pedestrian circulation, planting design, construction management
Michael Graves Associates, architecture
Hammel Associates, architecture

Columbia River Park
Client: Borough of Columbia
Principal consultants:
Derck & Edson Associates, master planning, site design, feasibility studies, community planning
Recreation and Park Solutions, recreation programming and operational analysis

Binns Park
Client: City of Lancaster
Principal consultants:
Derck & Edson Associates, master planning, landscape architecture, civil engineering, landscape design, construction observation
Hickey Architects and Tait Towers, stage design
Brinjac Engineering, lighting design
Sundance Water Design, fountain design
Greeneabum Structures, structural engineering
Organic Approach, organic maintenance
Art Metal Works, iron fence design

Culinary Institute of America Anton Plaza
Client: Culinary Institute of America
Principal consultants:
Derck & Edson Associates, site selection, feasibility, site design, green roof design
Noelker & Hull Associates, architects
American Hydrotech, Inc., green roof consultants
The Fountain People, fountain consultant
Hanover Architectural Products, paver system
Long Shadow, cast stone garden ornaments
Toro Irrigation, irrigation consultant

DUANY PLATER-ZYBERK & COMPANY

Seaside
Town Founder: Robert Davis
Principal consultants:
Duany Plater-Zyberk & Company, town planner
Deborah Berke, Carey McWhorter, Ernesto Buch, Walter Chatham, Tom Christ, Robert and Darryl Davis, Alex Gorlin, Steven Holl, Leon Krier, Rodolfo Machado and Jorge Silvetti, Scott Merrill, Robert Orr and Melanie Taylor, Aldo Rossi, Derrick

Smith, and Dan Solomon, key architectural designers
A. Douglas Duany, landscape design
www.seasidefl.com

Alys Beach
Town Founder: Gulf Coast Development, Jason Comer
Principal Consultants: Duany Plater-Zyberk & Company, town planner
Khoury-Vogt, town architect
Ana Alvarez, Juan Caruncho and Frank Martinez, Javier Cenicacelaya and Inigo Salona, DPZ & Co, Doug Farr, Michael Imber, Gary Justiss, Marieanne Khoury and Eric Vogt, Scott Merrill, Steve Mouzon, Robert Orr, Demetri Porphyrios, Julia Sanford, Derrick Smith, Eric Watson, key initial architectural designers
Glatting, Jackson, Kercher, Anglin, Lopez, Rinehart, transportation engineer
Moore Bass, civil engineer
Douglas Duany, landscape design
James Wassel, designer and illustrator
www.alysbeach.com

Schooner Bay
Town Founder: Lindroth Development Company, Orjan Lindroth
Principal consultants:
Duany Plater-Zyberk & Company, town planner
DPZ & Co, Michael Imber, Steve Mouzon, Andrew von Maur, Teofilo Victoria, Eric Watson, architecture and design
Keith Bishop, Michelle Duplaga-Bethel, James Moir, Kevin Main, Sean Hobbs, Scott Balcquiere, client consultant team
David Carrico, designer and illustrator

Sky
Town Founder: White Starr Development, Julia Starr Sanford and Bruce White
Principal consultants:
Duany Plater-Zyberk & Company, town planner
Joel Barkley, DPZ & Co, Jeff Dungan, Gary Justiss, Doug Luke, Oscar Machado, Steve Mouzon, Louis Nequette, Julia Sanford, Faisal Syed, architecture and design
Chance Powell, P.E. of Preble-Rish, Inc., civil engineering
J.R.'s Environmental Consulting, Kore Consulting, Florida State University's Center for Advanced Power Systems, The University of North Florida, environmental consultants and engineering
Zev Cohen & Associates, permitting and civil engineering
Mark David Major, project management
Dan Slone, code and legal
Teresa Baum, Louise Keim, sales/marketing and public relations
State of Florida Department of Environmental Protection, funding
James Wassell, designer and illustrator
www.skyflorida.net

East Fraserlands
Town Founder: Parklane Homes Ltd. and WesGroup Income Properties
Principal consultants:
Duany Plater-Zyberk & Company, town planner
James KM Cheng Architects Inc. architecture and urban design
Pat Pinnell, Joanna Alimanestianu, designers
Don Wouri Design, landscape design
GL Williams & Associates Ltd., Mark Holland of Holland Barrs, environmental and sustainability
www.city.vancouver.bc.ca/commsvcs/current-planning/current_projects/east_fraserlands/

Upper Rock
Town Founder: The JBG Companies
Principal consultants:
Duany Plater–Zyberk & Company, town planner
Burt Hill Kosar Rittelman Associates, sustainable design
Peter Katz, marketing
Maier Marketing, public relations
VIKA, planning support and civil engineering
Wells & Associates, transportation engineering
David Carrico, designer and illustrator
www.rockvillegateway.com

Marineland
Town Founder: Jacoby Development, Inc., Jim Jacoby
Principal consultants:
Duany Plater-Zyberk & Company, town planner
Oscar Machado, design
David Brain, consultant
Jono Miller, consultant
Daniel K. Slone, legal/codes
W. Terry Osborn, architecture
John Lambie, sustainable development
Dennis Creech, green building
University of Miami faculty: Rocco Ceo, Sonia Chao, Adib Curo and Carrie Penabad, Denis Hector, and Joanna Lombard, architecture and design
Tony Sease, Civitech; Chris Weddle, Otero Engineering, civil engineering
Steve Swann, Applied Technology & Management (ATM), environmental design
Mimi Vreeland, landscape design
Rick Hall, Hall Planning & Engineering, Inc., transportation engineering
James Wassell, designer and illustrator

EHRENKRANTZ ECKSTUT & KUHN ARCHITECTS

Battery Park City Master Plan
Client: Battery Park City Authority
Principal consultants:
Ehrenkrantz Eckstut & Kuhn Architects, master planning

Esplanade
Client: Battery Park City Authority
Principal consultants:
Ehrenkrantz Eckstut & Kuhn Architects, master planning
Olin Partnership, landscape architect

South Cove
Client: Battery Park City Authority
Principal consultants:
Ehrenkrantz Eckstut & Kuhn Architects, master planning
Stanton Eckstut, Mary Miss, and Susan Child, artists for South Cove Park

Liberty View
Client: Milstein Properties
Principal consultants:
Ehrenkrantz Eckstut & Kuhn Architects, architecture
Cosentini Associates, m/e/p engineer
Office of Irwin Cantor, structural engineer
Thomas Balsley Associates Landscape Architecture, landscape architect

Sites 23 and 24
Client: Milstein Properties
Principal consultants:
Ehrenkrantz Eckstut & Kuhn Architects, architecture
Costas Kondylis & Partners, architect of record
Gotham Construction, general contractor
Cosentini Associates, m/e/p engineer
WSP Cantor Seinuk, structural engineer
Langan Engineering, geotechnical engineer
Judith Heintz Landscape Architecture, landscape architect
Viridian Energy & Environmental, LLC

Arverne-by-the-Sea
Client: Benjamin-Beechwood LLC
Principal consultants:
Ehrenkrantz Eckstut & Kuhn Architects, master planning, architecture for Phase 1A housing
Quennell Rothschild & Partners, landscape architect
Vollmer Associates, civil engineer
Linnea Tillett Lighting Design, Inc., lighting design consultant
Sam Schwartz PLLC, transportation engineer
Ehasz Giacalone Architects, PC, architect of record

New River Las Olas
Client: Boca Developers
Principal consultants:
Ehrenkrantz Eckstut & Kuhn Architects, master planning, architectural design, public space design, entitlement/public review process consultation
Kimley Horn and Associates, civil engineer, landscape design
Kamm Consulting, m/e/p engineer
Blue Presentation, computer render

Gateway Center
Client: Catellus Development Corporation
Principal consultants:
Ehrenkrantz Eckstut & Kuhn Architects, master planning and architecture for Gateway Center
McLarand Vasquez & Partners, architects of MTA Building
The Olin Partnership, landscape architect
Fong & Associates, landscape architect
Rolf Jensen & Associates, life safety
Tamara Thomas, art program director
John A. Martin & Associates, structural engineer
Tsuchiyama & Kaino, mechanical engineer
Levine/Seegal Associates, electrical engineer
Mollenhauer, Higashi & Moore, civil engineer
Howard Brandston & Partners, lighting consultant
Nigel Nixon & Partners, paving consultant
Sussman Prejza & Company, graphic consultant
Charles Pankow Builders, contractor

Hollywood & Highland
Client: TrizecHahn
Principal consultants:
Ehrenkrantz Eckstut & Kuhn Architects, master planning and architecture
Altoon and Porter, architect of record
Rockwell Group, theater architect
Wimberly Allison Tong & Goo, hotel architect
Rios Associates, landscape architect
Robert Englekirk Consulting Structural Engineers, structural engineer
Levine/Seegel Associates, m/e/p engineer
Fisher Marantz Renfro Stone, theater lighting
Lighting Design Alliance, awards walk lighting
Sussman/Prejza & Company, graphics
Rolf Jensen & Associates, fire protection
Lerch Bates North America, transportation
McCarthy Building Companies, general contractor

HUGHES, GOOD, O'LEARY & RYAN, INC.

Terminus
Client: Cousins Properties
Principal consultants:
Hughes, Good, O'Leary & Ryan, Inc., urban design, master planning, leading two-day charrette involving architects, engineers, and client
Duda/Paine Architects, LLP, design architect
HKS, Inc., architect of record

Magnolia
Client: Magnolia Development LLC
Principal consultants:
Hughes, Good, O'Leary & Ryan, Inc., urban design analysis and planning
Shook Kelley, Inc., perception design firm

Allen Plaza
Client: Barry Real Estate Companies
Principal consultants:
Hughes, Good, O'Leary & Ryan, Inc., urban design and planning, public space design including plazas and streetscapes
Pickard Chilton Architects, architect

Columbus Riverfront Office Building
Client: W.C. Bradley Companies
Principal consultants:
Hughes, Good, O'Leary & Ryan, Inc., urban planning and design of open spaces
Crawford McWilliams Hatcher Architects, Inc., architect

JAMES, HARWICK + PARTNERS, INC.

Cityville Greenville
Client: FirstWorthing Co.
Principal consultants:
James Harwick + Partners, Inc., planning, architectural design
RTKL, landscape architecture

The Commons at Atlantic Station
Client: The Lane Companies
Principal consultants:
James Harwick + Partners, Inc., architects
Garden Architects, Inc., landscape architect

Cityville Southwest Medical District
Client: FirstWorthing Co.
Principal consultants:
James Harwick + Partners, Inc., planning, architecture
RTKL, landscape architecture

West Highlands (Columbia Crest, Columbia Estates, and Columbia Heritage Senior Residences)
Client: Columbia Residential
Principal consultants:
James Harwick + Partners, Inc., architectural design
Site Solutions and SGN&A, landscape architect

Museum Place
Client: JaGee Real Properties, L.P., Townsite Company
Principal consultants:
James Harwick + Partners, Inc., planning and urban design

Cityville Fitzhugh
Client: FirstWorthing Co.
Principal consultants:
James Harwick + Partners, Inc., architectural design
RTKL, landscape architects

JOSEPH WONG DESIGN ASSOCIATES

16th/Market Affordable Residential Project
Client: JMI Realty, Lennar, Father Joe's Village (residential nonprofit)
Principal consultants:
Joseph Wong Design Associates, architectural design
Burkett & Wong, civil & structural
Garbini & Garbini, landscape

North County Regional Education Center
Client: San Marcos Unified School District and the San Diego County Office of Education
Principal consultants:
Joseph Wong Design Associates, architectural design
Ron Teshima, landscape architects
Flores Lund Consultants, structural engineers
RBF Consulting, civil engineers
HVAC Engineering, plumbing & mechanical
ILA Zammit, electrical engineers

Lane Field
Client: Port of San Diego
Principal consultants:
Joseph Wong Design Associates, lead architect
Davis Davis Architects, design/associate architect
Hargreaves Associates, landscape architect
Project Design Consultants, civil engineer

Sofitel JJ Oriental Hotel
Client: Jin Jiang Group
Principal consultants:
Joseph Wong Design Associates, architectural design
Shanghai Architectural & Engineering Design Institute, architect of record

Deep Blue Plaza
Client: Greentown Development Group
Principal consultants:
Joseph Wong Design Associates, architectural design
EDAW, landscape architect
BLD & Steve Leung, interiors

JPRA

The Village of Rochester Hills
Client: Robert B. Aikens & Associates, LLC
Principal consultants:
JPRA Architects, master planning, construction documents, field administration, lighting design
Grissim Metz Andriese Associates, landscape architecture

Somerset Collection South Renovations
Client: The Forbes Company
Principal consultants:
JPRA Architects, architectural design, construction documents, field administration
Focus Lighting, lighting
Wet Design, fountain design

Palladium
Client: The Palladium Company
Principal consultants:
JPRA Architects, master planning, architecture, construction documents, field administration, graphics, interiors
Grissim Metz Andriese Associates, landscape architecture

LANDDESIGN

SouthPark Mall
Client: Simon Properties and Faison
Principal consultants:
LandDesign, landscape architecture, civil engineering
Bartlett Associates Megastrategies, architecture

The Towne of Seahaven
Client: The Seahaven Companies; Seahaven Development, Inc.
Principal consultants:
LandDesign, master planning, urban design, landscape architecture
Baker Barrios Architects, project residential architect
Intrawest U.S. Property Management, Inc., marketing and management consultant
Lovelace Interiors, interior design

Oz Architecture, village retail architect
PBS&J, civil engineer

Downtown Silver Spring
Client: PFA Silver Spring, LLC
Principal consultants:
LandDesign, landscape architecture, construction documents, construction coordination
Maryland National Capital Parks and Planning Commission
Loiederman Soltesz Associates, civil engineer
Brown, Craig, Turner, architects
WDG Architecture, architects
Stern and Associates, Inc., electrical engineer for private work
Richter and Associates, public utilities
Century Engineering, structural engineer
Wells and Associates, Inc., traffic engineer
Lynch & Associates, Inc. and Hydro Designs, Inc., irrigation
Wesco Fountains, fountain construction
Artbridge, mosaic tile artist
Montgomery County, public sector client

Shenzhen Bay Seafront Urban Design
Client: Government of Shenzhen, Guangdong Province, P.R. China
Principal consultants:
LandDesign, urban design, landscape architecture

Cool Springs Mixed-Use Development
Client: Crescent Resources
Principal consultants: LandDesign, master planning and design

LESSARD GROUP

Midtown Reston Town Center
Client: KSI Services
Principal consultants:
Lessard Group, feasibility, architectural design, construction administration
Tadjer-Cohen Edelson Associates, Inc., structural
Bovis Lend Lease, contractor
Cherry Lane Electrical Services, Inc., electrical
Lewis Scully Gionet, Inc., landscape
Urban Engineering & Associates, Inc., civil
Forrest & Perkins, interiors
C.M. Kling Associates, lighting

Clarendon Park
Client: Eakin/Youngentob Associates, Inc., developer
Principal consultants:
Lessard Group, feasibility, land planning, architectural design

Bowman Consulting Group, civil engineer
Parker Rodriguez, landscape architecture
Alliance, structural engineer

Port Imperial
Client: Roseland Property Co.
Principal consultants:
Lessard Group, feasibility, architectural design

Chatham Square
Client: Samuel Madden Homes
Client for ARHA units: Midcity Development and the City of Alexandria Residential Housing Authority
Principal consultants:
Lessard Group, feasibility, architectural design, permit and construction documents and construction administration
EYA, general contractor
Bowman Consulting Group, civil engineer
Studio 39, landscape architect
Alliance Structural Engineers, structural engineer
META Engineers, mep engineer

LUCIEN LAGRANGE ARCHITECTS

X/O Condominiums, 1712 South Prairie Avenue
Client: Frankel & Giles
Principal consultants:
Lucien Lagrange Architects, architecture
Halvorson and Partners, structural engineer
Cosentini Associates, Inc., mep engineer
Eriksson Engineering Associates, Ltd., civil engineer
Site Design Group, Inc., landscape architect
H2O Design Group, aquatic engineer

840 North Lake Shore Drive
Client: LR Development Company
Principal consultants:
Lucien Lagrange Architects, architecture
Thornton Tomasetti Engineers, structural engineer
Environmental Systems Design, electrical and mep engineer
James McHugh Construction Company, general contractor
Wolff-Clements and Associates, landscape architect
Gate Precast Company, precaster

Hard Rock Hotel
Client: Mark IV Realty Group
Principal consultants:
Lucien Lagrange Architects, architecture
Wiss Janney Elstner Associates, Inc., exterior restoration consultant
EME, LLC, mep engineer
Tylk Gustafson Reckers Wilson Andrews, LLC,

structural engineer
Yabu Pushelberg, interior design
Pepper Construction, general contractor

Park Kingsbury
Client: Cataldo Family Enterprises
Principal consultants:
Lucien Lagrange Architects, architecture
Halvorson Partners, structural engineer
Melvin Cohen & Associates, mep engineer

MBH

Sonoma Mountain Village
Client: Codding Enterprises & St. James Properties
Principal consultants:
MBH Architects, architect
Allen Land Design, landscaping
Balance Hydrologics, Inc., civil engineering & habitat consulting
BFK Engineers, civil engineers/surveyors
Carlenzoli & Associates, civil surveyor
Farrell, Faber & Associates, homes architecture
Fisher & Hall Urban Design, urban planner
Lois Fisher/Laura Hall, urban planner
Gauger & Associates Advertising, advertising/marketing
Kema Green, green building
Del Starrett Architecture, commercial architecture
Wix Architecture, lofts/condominiums/apartments architecture

Broadway Arms
Client: CIM Group
Principal consultants:
MBH Architects, architectural design of commercial architecture and lofts/condominiums/apartments
Melendrez Design Partners, landscaping
Huitt-Zollars, Inc., civil engineers/surveyors/planners
International Parking Design, parking consultant
Lighting Design Alliance, lighting designer
Lois Fisher / Laura Hall, urban planner

Broadway Grand
Owner: Signature Properties
Principal consultants:
MBH Architects, architect
Signature Properties, contractor
Nishkian Menninger, structural engineer
BKF Civil Engineers, civil engineer
Critchfield Mechanical Inc., mechanical
Cupertino Electric Inc., electrical
W.L. Hickley Sons, Inc., plumbing
Superior Automatic Sprinkler Company, fire protection

Charles M. Salter Associates, Inc., acoustical engineer
Guzzardo Partnership, landscape
Treadwell & Rolo, geotechnical engineer
Simpson, Gumpertz & Heger, waterproofing

Odeon Union Square – 150 Powell Street
Client: Union Property Capital
Principal consultants:
MBH Architects, architecture, construction administration
MCA, structural
Ajmani & Pamini, mep
Luk & Associates, civil
Salter & Associates, acoustical
McCloudesich, fountain

MOULE & POLYZOIDES

Robert Redford Building for the Natural Resources Defense Council
Client: Natural Resources Defense Council (NRDC)
Principal consultants:
Elizabeth Moule & Stefanos Polyzoides, feasibility, concept design, schematic design through construction documents, construction administration
Syska and Hennessy, mep
Environmental Planning and Design, water recycling
CTC Energentics, LEED certification

Del Mar Station
Client: originally, Urban Partners, project sold during construction to Archstone-Smith
Principal consultants:
Elizabeth Moule & Stefanos Polyzoides, feasibility, concept design, schematic design through design development
Nadel Partners, executive architect
Melendrez Design, landscape architect

Rio Nuevo
Client: Rio Development Company
Principal consultants:
Elizabeth Moule & Stefanos Polyzoides, architecture and urbanism
Charrette Leader
Oscar Machado, urbanism
Paul Weiner, Architect, DesignBuild Collaborative, architecture
The WLB Group, engineering
Ann Philips, landscape and water harvesting

MSI

Crocker Park
Client: Stark Enterprises
Principal consultants:
MSI, schematic site design through construction observation for all site improvements and related landscape architectural features
Bialosky & Partners, architecture
HLB Lighting Design, lighting design

Celebration Hotel
Client: The Kessler Enterprise, Inc.
Principal consultants:
MSI, hardscape, landscape, irrigation
Gund Partnership, design architect
Lindsay Pope Brayfield & Associates, Inc., architect of record
PBS&J, civil engineer

North Bank Park
Client: City of Columbus Recreation and Parks Department
Principal consultants:
MSI, project planning and landscape architecture
Acock Associates Architects, architecture
HKI Associates Inc., architect of record
Miles McClellan, construction management
EMH&T, civil engineering
Parsons Transportation, transportation engineering
Envirotech, environmental consulting
Burgess & Niple, river and floodway engineering
Kolar Design, graphic design

Nationwide Arena District
Client: Nationwide Realty Investors
Principal consultants:
MSI, master planning and landscape architecture
360 Architecture, architecture
Acock Associates Architects, architecture
Lupton Rausch Architecture + Interior Design, architecture
Meleca Architecture, architecture
NBBJ, architecture
EMH&T, civil engineering
Parsons Transportations, transportation engineering
Kolar Design, graphic design

MULVANNYG2 ARCHITECTURE

Redmond City Hall
Client: Wright Runstad & Company (owner) and City of Redmond (public agency) in public / private partnership
Principal consultants: Mulvanny G2 Architecture, full architectural design, conceptual design, schematic design, design development, construc-

tion documents, construction administration, public space interior design
Lease Crutcher Lewis, general contractor
Magnusson Klemencic Associates, civil and structural
Geo Engineers, geotechnical
Transpo Group, transportation
Hewitt Architects, landscape
MacDonald-Miller Facility Solutions, mechanical and plumbing
Roth Hill Engineers, surveyor
Sasco, electrical
Perrault Interiors, office area interior design

Zhejiang Fortune Financial Center
Client: Zhejiang Telong Real Estate Development Co. Ltd.
Principal consultants: Mulvanny G2 Architecture, master planning, conceptual design, schematic design

Zhangjiang Semiconductor Research Park, Phase II
Client: Shanghai Zhangjiang Semiconductor Park Development Co., Ltd.
Principal consultants: Mulvanny G2 Architecture, master planning, design architect
ECADI, Shanghai (East China Architectural Design and Research Institute), local project architect, structural/mechanical/electric engineer

Fudan Crowne Plaza Hotel
Client: Shanghai Shangtou Investment Group Co. Ltd.
Principal consultants: Mulvanny G2 Architecture, design architect (conceptual design, schematic design, design development)
Shanghai Architectural Design & Research Institute, local project architect, structural/mechanical/plumbing/electricity engineer
Fleming Landscape, landscape
GIFL, interior design
Zhangiang Long Yuan Construction Co., general contractor

NEWMAN GARRISON GILMOUR + PARTNERS

Watermarke
Client: Sares-Regis Group
Principal consultants:
Newman Garrison Gilmour + Partners, planning and architectural design
Lifescapes International, landscape
Catalina Group, model interiors
Style Interiors, clubhouse interiors
Regis Contractors, builders
Dale Christian Engineers, structural
Parks Mechanical, plumbing
LDI Mechanical, mechanical
DGM & Associates, electrical
Parkitects, garage

Avalon Del Rey
Client: Avalon Bay Communities
Principal consultants:
Newman Garrison Gilmour + Partners, planning and architectural design
Dale Christian Engineers, structural
Parkitects, garage
Lifescapes International, landscape
Hall & Foreman, civil
DGM & Associates, electrical
Hart Laboratories, environmental engineering
Faulkner Design Group, interiors
LDI mechanical, mechanical
Parks Mechanical, plumbing

Paxton Walk
Client: Blue Marble Development
Principal consultants:
Newman Garrison Gilmour + Partners, planning and architectural design
Lifescapes International, landscaping
Style Interiors, interiors
Flack & Kurtz, mep
Veneklasen Associates, acoustical
Fource Coimmunications, Ltd., signage
M.E. Nollkamper & Associates, utility consultants
Martin & Peltyn, Inc., structural
Carter & Burgess, Inc., structural
The Kelly Group, structural
Wrought Engineering, structural
Enaleman & Associates, ADA consultant
HKA Elevator Consulting, Inc., elevator
La Jolla Pacific, Ltd., waterproofing
Parkitects, parking garage

Amerige Pointe
Client: The Morgan Group
Principal consultants:
Newman Garrison Gilmour + Partners, planning and architectural design
Vandorpe Chou Associates, structural
IMA Design, landscape
Faulkner Design Group, interior design
KHR Associates, civil engineers
AMPAM/LDI Mechanical, mechanical and plumbing
Donn C. Gilmore Asociates, electrical

O'BRIEN & ASSOCIATES

City Lights
Client: Margaux Properties
Principal consultants:
O'Brien & Associates, architectural services

The Harbor
Client: Rob and Sarah Whittle
Principal consultants:
O'Brien & Associates, design and architectural services
TBG Landscape Architects, hardscape and landscape

Highland Village
Client: Regency Centers
Principal consultants:
O'Brien & Associates, architectural services
Linda Tycher and Associates, hardscape and landscape

Sugarland Town Center Expansion
Client: Sugarland Properties
Principal consultants:
O'Brien & Associates, architectural services

PERKOWITZ+RUTH ARCHITECTS/STUDIO ONE ELEVEN

Court Street West Specific Plan
Client: LNR, Manchester Group, The Street Company
Principal consultants:
Perkowitz+Ruth, Studio 111, concept design, specific plan
FORMA, specific plan, landscape
DRC, civil
IBI, traffic/parking

Galerias Hipodromo
Client: Gruppo Fema
Principal consultants:
Perkowitz+Ruth Architects, architectural design
The Collaborative West, landscape/hardscape
Lighting Design Alliance, lighting
Parking Design Group, parking consultant
Wiseman & Rhoy, structural
Bridur Construcciones, S.A.D.C.V., civil engineers

Downtown Long Beach Visioning
Client: City of Long Beach
Principal consultants:
Perkowitz+Ruth, Studio 111, urban design, three-dimensional modeling and animated fly-throughs, zoning consultation
Meyer Mohaddes Associates, traffic consultant,

Bella Terra
Client: DJM Capital Partners, J.H. Snyder Co., The Ezralow Company
Principal consultants:
Perkowitz+Ruth Architects, architectural design and planning
LA Group, landscape
Lighting Design Alliance, lighting
Romero Thorsen Design, signage design
Davar and Associates, mechanical engineer
GLP Engineering, Inc., electrical engineer
Wiseman + Rohy Structural Engineers, structural engineer
EN Engineers, civil engineer
Jerde Partnership, Inc., early design concepts

The Garden
Client: Pao Huei Construction Company, Ltd.
Principal consultants:
Perkowitz+Ruth, Studio 111, building envelope design
EPT, landscape design

Melrose Triangle
Client: Charles Company
Principal consultants:
Perkowitz+Ruth, Studio 111, architectural design
Saiful Bouquet Structural Engineers, Inc., structural
Rios Clemente Hale Studios, landscape
SGH Design + Consulting Engineers, waterproofing
Newsom Design, environmental and graphic design
Patrick B. Quigley Associates, lighting

Bridge Street Town Centre
Client: O & S Holding, LLC
Principal consultants:
Perkowitz+Ruth Architects, architectural design and planning; currently in schematic design phase of retail and theatre component
Francis Krahe & Associates, lighting
RTKL, environmental graphics
HRP Landesign, landscape
TRC International Ltd., structural engineer
Breen Engineering, mechanical/electrical engineer

RLC ARCHITECTS

Aqua Vista Lofts
Client: Sharpe Project Developments and Patron Development
Principal consultants:
RLC Architects, schematic design, design development, construction documents and construction administration
Johnson Structural Group, structural
Puga and Associates, mep
F&R Fire Protection, fire protection
Ecoplan, landscape

U.S. Epperson/Lynn Insurance Group Corporate Headquarters
Client: U.S. Epperson/Lynn Insurance Group
Principal consultants:
RLC Architects, site feasibility, schematic design, design development, construction documents and construction administration
O'Donnell Nacarrato Mignogna Jackson, structural engineer
TLC Engineering for Architecture, mep engineer
Ecoplan, landscape

951 Yamato
Client: Patriot Realty
Principal consultants:
RLC Architects, design development, construction documents and construction administration
Burton Braswell Middlebrooks, structural
Thompson Youngross Consulting Engineers, mep

Domus Office Tower
Client: Domus Development Group
Principal consultants:
RLC Architects, site selection, schematic design, design development, construction documents and construction administration
Donnell Duquesne Albaisa Engineers, structural

RTKL ASSOCIATES

New Jiang Wan Cultural Center
Client: Shanghai Chengtou City Land Group
Principal consultants:
RTKL Associates, architecture
Shanghai Institute Architectural Design and Research Company, associated architects, MEP, structural engineers,
fire protection services
Shanghai Construction Company, general contractor

The China Film Museum
Client: Chinese Film Museum Project Committee
Principal consultants:
RTKL Associates, architecture
Beijing Institute of Architectural Design and Research, local architect

Highmark Data Center
Client: Highmark, Inc.
Principal consultants:
RTKL Associates, architecture, MEP engineering, structural engineering, interior design, special systems design
CS Technology, IT consultant
E. Howard Black and Associates, civil engineering
Cerami & Associates, Inc., acoustical
CMS Innovative Solutions, audio-visual
Hughes Associates, life safety/fire protection,
Terralogos, sustainability/LEED certification,
Paul Waddelove & Associates, cost estimating

Reginald F. Lewis Museum of Maryland African American History and Culture
Client: Maryland African American Museum Corporation
Principal consultants:
Freelon/RTKL A Joint Venture, architecture, environmental graphic design, MEP and structural engineering, interior architecture
Whiting-Turner Contracting, general contractor
Bayview Landscaping, Inc., landscape architecture
Design and Integration, audio-visual design

City Crossing
Client: China Resources (Shenzhen) Co., Ltd.
Principal consultants:
RTKL Associates, master planning, architecture, environmental graphic design
The Architectural Design & Research Institute of Guangdong Province, local architect
Place Planning & Design, landscape design consultant
Kaplan Partners Architectural Lighting, lighting consultant
Flack + Kurtz Consulting Engineers, electrical and mechanical engineering consultant
Rolf Jensen & Associates, Inc., fire protection and safety consultant
Hesselberg, Keesee & Associates, Inc., vertical transportation consultant
Curtain Wall Design & Consulting, Inc., curtain wall design consultant
CIECC Engineering & Construction Project Management Corp., supervising engineer

SASAKI ASSOCIATES

DART CBD Transit Mall
Client: Dallas Area Rapid Transit Authority
Principal consultants:
Sasaki Associates, planning, urban design, landscape architecture, civil engineering
Arredondo Bruz & Associates, Inc., civil engineering
Barton-Aschman, Inc., traffic
Berryhill-Loyd Associates, Inc., structural engineering
H.M. Brandston & Partners, Inc., lighting
Campos Engineering, mep engineering
Brad Goldberg, artist
Haywood, Jordon, McCowan, Inc., architect

Huitt-Zollars, Inc., civil engineering
Oglesby Group, architect

Providence 2020
Client: City of Providence, Rhode Island
Principal consultants:
Sasaki Associates, planning, urban design, landscape architecture
Vanasse Hangen Brustlin, infrastructure engineering
ZHA, Inc., economic consultant
Barbara Sokoloff Associates, Inc., community development planning

Schenley Plaza
Client: City of Pittsburgh, Department of City Planning
Principal consultants:
Sasaki Associates, landscape architecture
Environmental Planning and Design, LLC, local landscape architect
Quad 3 Group, mep engineering
Brandston Partnership Inc., lighting
The Gateway Engineers, survey
LeMessurier Consultants, structural engineering

New Jersey Urban Parks Master Plan Competition
Client: Division of Property Management and Construction
Principal consultants:
Sasaki Associates, planning, urban design, landscape architecture
Center for Urban Restoration Ecology, ecological restoration
Cultural Resource Consulting Group, cultural resource management
TRC Omni Environmental Corporation, environmental
Janet Echelman, Inc., artist

SMITHGROUP

Downtown Detroit YMCA
Client: YMCA of Metropolitan Detroit
Principal consultants:
SmithGroup, architecture and engineering
Fisher Dachs Associates, theater design
Water Technology, Inc., pool design
Tucker Young Jackson Tull, Inc., civil engineering

Discovery Communications Headquarters
Client: Discovery Communications
Principal consultants:
SmithGroup, architectural design
ECS, Inc., geotechnical
Kimley-Horn & Associates, Inc., traffic management
VIKA Incorporated, civil engineering

Flack + Kurtz Consulting Engineers, mechanical engineering
KTLH Consulting Engineers, structural engineers
Walker Parking Consultants, parking consultant
EDAW, Inc., landscape architect
Richter & Associates, utility management
Gensler, interiors
Schabel Engineering Associates, geotechnical and inspection

Visteon Village Corporate Headquarters
Client: Visteon Corporation
Principal consultants:
SmithGroup, architecture, engineering, landscape architecture
Lovett Consultants, building code consultant
InkSpot Design, Inc., wayfinding signage
Gieger & Hamme, Inc., acoustical consultant
Technical Inspection, Inc., elevator consultant
Rowan Williams Davies & Irwin, Inc., snow drift/wind
Cerami & Associates, audio/visual consultant
E.F. Whitney, Inc., food services consultant
NTH Consultants, Ltd., geotechnical consultant
J.E. Edinger Associates, Inc., liminological consultant
Somat Engineering, geotechnical consultant
Door Security Solutions, door hardware consultant
Zoyes East, model building

STARCK ARCHITECTURE AND PLANNING

Beacon Point at Liberty Station
Client: McMillin Companies
Principal consultants:
Starck Architecture + Planning, schematic site design, preliminary design, construction documents
M.W. Steele Group, planner
RBF Consulting, civil engineer
Swanson and Associates, structural engineer
Design Line Interiors, interior design

Portico
Client: Sandcor Harborside, LLC
Principal consultants:
Starck Architecture + Planning, schematic site design, preliminary design, construction documents
Ledcor SD Construction, contractor
Brukett & Wong, structural engineer
Fuscoe Engineering, civil engineer
URS, acoustical engineer
Jim Mann & Associates, mechanical engineer
Nutter Electrical Design, electrical engineer
John Hanna Landscape Architecture, landscape architect blue motif, interior design

Bridges at Escala
Client: D.R. Horton
Principal consultants:
Starck Architecture + Planning, site plan review, preliminary design, construction documents
Hortowitz, Taylor, Kushkaki Engineering, structural engineer
Project Design Consultants, civil engineer
Gillespie Moody Patterson, Inc., landscape architect
CH Design Group, interior design

Ponto Beachfront Village
Client: K. Hovnanian Companies
Principal consultants:
Starck Architecture + Planning, schematic site design, preliminary design, later phases in progress
Howes Weiler & Associates, land use consultant
Project Design Consultants, civil engineer
Gillespie Moody Patterson, Inc., landscape architect

STUDIO 39

Workhouse Art Center at Lorton
Client: Lorton Arts Foundation
Principal consultants:
Studio 39, landscape architecture
Hellmuth, Obata + Kassabaum, P.C., architecture
Greenhorne & O'Mara, civil engineering

Aurora Condominiums
Client: RST Development
Principal consultants:
Studio 39, landscape architecture
A.R. Meyers & Associates Architects, Inc., architecture
VIKA, Inc., civil engineering

WRIT Rosslyn Center
Client: Washington Real Estate Investment Trust
Principal consultants:
Studio 39, landscape architecture
Architects Collaborative, architecture
VIKA, Inc., civil engineering

Ford's Landing
Client: EYA Development
Principal consultants:
Studio 39, landscape architecture
Lessard Group, planning and architecture
Bowman Consulting Group, civil engineering

TORTI GALLAS AND PARTNERS

City Vista
Client: L5K, LLC, a joint venture of Lowe Enterprises Real Estate Group, CIM Group, Bundy Development Corporation and The Neighborhood Development Company
Principal consultants:
Torti Gallas and Partners, planning, architecture, urban design, construction phase services
Michael Marshall Architects, associate architect
Delon Hampton & Associates, civil
Smislova Kehnemui & Associates, structural
GHT Limited, mep
ECS, Limited, geotechnical
Gorove/Slade Associates, transportation/traffic
Lee & Associates, landscape architect
Hartman Design Group, interior design

The Ellington
Client: Donatelli & Klein
Principal consultants:
Torti Gallas and Partners, planning, architecture, urban design, construction phase services
Donohoe Construction Company, builder
Bowman Consulting Group, civil
Schnabel Engineering Associates, geotechnical
Tadjer Cohen Edelson Associates, structural
Schwartz Engineering, mep
Land Design, Inc., landscape architect
Design Works Interiors, interior design

3030 Clarendon Boulevard
Client: Saul Centers, Inc.
Principal consultants:
Torti Gallas and Partners, planning, architecture, urban design, construction phase services
Bowman Consulting Group, civil
Tadjer Cohen Edelson Associates, structural
GHT Limited, mep
Michael V. Bartlett, Inc., landscape architect

Highland Park
Client: Donatelli & Klein
Principal consultants:
Torti Gallas and Partners, planning, architecture, urban design, construction phase services
Bowman Consulting Group, civil
Cates Engineering, structural
GHT Limited, mep
Froehling & Robertson, geotechnical

Kenyon Square
Client: Donatelli & Klein
Principal consultants:
Torti Gallas and Partners, planning, architecture, urban design, construction phase services
Bowman Consulting Group, civil
Cates Engineering, structural
GHT Limited, mep
Froehling & Robertson, geotechnical

Park Triangle
Client: Triangle Development Associates, LLC
Principal consultants:
Torti Gallas and Partners, planning, architecture, urban design, construction phase services
Loiederman Soltesz Associates, civil
Smislova Kehnemui & Associates, structural
Metropolitan Engineering Shapiro-O'Brien, mep
Mueser Rutledge Consulting Engineers, geotechnical
Parker Rodriguez, Inc., landscape architect
Forma Design, interior design

Park Place Condominiums
Client: Donatelli & Klein
Principal consultants:
Torti Gallas and Partners, planning, architecture, urban design, construction phase services
A. Morton Thomas & Associates, civil
Froehling & Robertson, geotechnical
GHT, Limited, mep
Cates Engineering, structural
Parker Rodriguez, Inc. landscape architect
Hickok Cole Architects, interior design

Twinbrook Commons
Client: The JBG Companies
Principal consultants:
Torti Gallas and Partners, planning, architecture, urban design
VIKA, Inc., civil
Wells & Associates, transportation
Walker Parking Consultants, parking

Shirlington Village
Client: Bozzuto Development Company
Principal consultants:
Torti Gallas and Partners, planning, architecture, urban design, construction phase services
Dewberry & Davis, LLC, civil
ECS, Limited, geotechnical
Green Industries and Petrossian & Associates, mep
Smislova Kehnemui & Associates, structural
Hartman Design Group, interior design

The Residences at the Greene
Client: Steiner & Associates
   (Partner: Mall Properties)
Principal consultants:
Torti Gallas and Partners, residential architecture
Design Development Group, project master planner and designer
Meacham & Apel Architects, architect of record
Messer Construction, general contractor
Jezerinac-Geers & Associates, Inc. structural
M Retail, mep

Cooper's Crossing
Client: Steiner & Associates
Principal consultants:
Torti Gallas and Partners, planning, urban design, programming
Urban Design Associates, associate architect
Copley Wolff Design Group, landscape architect
Cooper's Ferry Development Association, other consultant
Michael Vergason Landscape Architects, residential landscape designer
The Edge Group, landscape architect

Lansdowne Village Greens
Client: Lansdowne Town Center, LLC
Principal consultants:
Torti Gallas and Partners, concept design, entitlement, design code
Christopher Consultants and Bowman Consulting, civil
Reed Smith, attorney
Lewis Scully Gionet, landscape architect
Michael Morrissey, renderer

Fort Belvoir Military Housing
Client: Department of the Army/Clark Pinnacle, LLC
Principal consultants:
Torti Gallas and Partners, master planning, neighborhood planning, architectural design, construction phase services
Parker Rodriguez, Inc., landscape architect
Greenhorne & O'Mara, civil
Cates Engineering, structural
JCS Engineering, mep
Aida Color, color consultant

Martin Luther King Plaza
Client: Philadelphia Housing Authority
Principal consultants:
Torti Gallas and Partners, master planning, architectural design, urban design, construction phase services
Domus, Inc., builder
Brickman Group, landscape archtiect

Centergate at Baldwin Park
Client: Pritzker Residential
Principal consultants:
Torti Gallas and Partners, master planning, architectural design, urban design, construction phase services
PBS&J, civil
Danny Powell Landscape Architect, landscape architect

KTD Consultants, mep
R.L. Plowfield & Associates, structural

Salishan
Client: Tacoma Housing Authority, owner; Lorig Associates, developer
Principal consultants:
Torti Gallas and Partners, planning, architecture, urban design, construction phase services
McGranahan Architects and Environmental Works
Community Design Center, associate architects
The Berger Partnership, landscape architect
Parametrix, civil
Putnam Collins Scott Associates, structural
Key Engineering, plumbing
Active Engineering, electrical
Walsh Construction Co./WA, builder

Vantaggio Baldwin Hills
Client: MSA Development
Principal consultants:
Torti Gallas and Partners, planning, architecture, urban design, construction phase services
S.Y. Lee Associates, mep engineer
Englekirk Partners, structural engineer
Fong Hart Schneider + Partners, landscape architect
Sukow Engineering, civil

Downtown Upland
Client: CIM Group
Principal consultants:
Torti Gallas and Partners, master planning, urban design
Mobility Group, traffic consultant

Taylor Yards
Client: McCormack Baron Salazar
Principal consultants:
Torti Gallas and Partners, master planning, architectural design, urban design, neighborhood planning

TVS

Washington Convention Center
Client: The Washington Convention Center Authority
Principal consultants:
TVS, Devrouax & Purnell, Mariani Architects Engineers, TVS Interiors
John J. Christie & Associates, henry Adams, Inc., James Madison Cutts, Ross Bryan Associates, Daniels and Associates, Mueser Rutledge Consulting Engineers, Jackson & Tull, Clark/Smoot, Joint Venture, JBG, HNTB and Turner Construction

Technology Square, Georgia Institute of Technology
Client: Georgia Institute of Technology and The Georgia Tech Foundation
Principal consultants:
TVS, architects
TVS Interiors, interiors
Holder/Hardin, general contractor
W P Moore, structural engineer
Newcomb & Boyd, mechanical engineer
Jones Lang LaSalle, project managers

Plaza Norte
Client: Mall Plaza
Principal consultants:
TVS, architects
Jaime Vargas C., production architect
TVS Interiors, interiors
Arnello & Viveros Arquitectos, landscape architect
Salfacorp, general contractor
Mall Plaza, developer
PBS Arquitectos, lighting design

Short Pump Town Center
Client: a joint venture of Forest City Development and Pruitt Associates
Principal consultants:
Forest City Enterprises and MJGT Associates, owner
TVS, design architect
TVS, interior design
KA Architecture, Inc., associate architect
Barton Associates, Inc., mechanical/electrical
Shenberger and Associates, structural
Whiting-Turner Contracting Co., contractor
Roy Ashley and Associates, landscape architect
Bliss Fasman, Inc., lighting design
Huie Design, graphic design
McKinney and Co., civil

Triangle Town Center
Client: The Richard E. Jacobs Group
Principal consultants:
TVS, architects
TVS Interiors, interiors
T. Kondos & Associates, lighting designer
MESA Design, landscape architect
Hoar Construction, general contractor

Georgia Aquarium
Client: Marcus Foundation
Principal consultants:
TVS, TVS Interiors, The Jerde Partnership International, Inc., PGAV, Inc., Heery International, Inc., Syska Hennessy Group, Uzun & Case, Brasfield & Gorrie

Index by Project

16th/Market Affordable Residential Project, San Diego, CA, **130**
3949 Lindell Boulevard, St. Louis, MO, **52**
840 North Lake Shore Drive, Chicago, IL, **164**
951 Yamato, Boca Raton, FL, **230**

Alexan Pacific Grove, Orange, CA, **40**
Allen Plaza, Atlanta, GA, **116**
Amerige Pointe, Fullerton, CA, **208**
The Architect Collection at Stapleton, Denver, CO, **70**
Aqua Vista Lofts, Fort Lauderdale, FL, **226**
Artessa at Quarry Village, San Antonio, TX, **54**
Aurora Condominiums, Silver Spring, MD, **168**
Arverne-by-the-Sea, Arverne, NY, **108**
Avalon Del Rey, Los Angeles, CA, **204**

Battery Park City, New York, NY, **106**
Beacon Point at Liberty Station, San Diego, CA, **258**
Beijing CBD Office Park, Beijing, China, **32**
Beijing Century City East, Beijing, China, **30**
Bella Terra, Huntington Beach, CA, **221**
Belle Creek, Henderson, CO, **66**
Belmar Row Houses, Lakewood, CO, **68**
Binns Park, Lancaster, PA, **93**
Boca Raton Mixed-use, Boca Raton, FL, **80**
Bridge Street Town Centre, McKinney, TX, **224**
Bridges at Escala, San Diego, CA, **261**
Broadway Arms, Anaheim, CA, **172**
Broadway Grand, Oakland, CA, **173**
The Brownstone Collection at Highlands Ranch Town Center, Highlands Ranch, CO, **69**

Celebration Hotel, Celebration, FL, **188**
Chatham Square, Alexandria, VA, **160**
The China Film Institute, Beijing, China, **236**
China Science and Technology Museum, Beijing, China, **28**
Circuit Court of Cook County, Chicago, IL, **58**
City Center at Oyster Point, Newport News, VA, **86**
City Crossing, Shenzhen, China, **240**
City Lights, Dallas, TX, **210**
Cityville Fitzhugh, Dallas, TX, **128**
Cityville Greenville, Dallas, TX, **122**
Cityville Southwestern Medical District, Dallas, TX, **125**
Clarendon Park, Arlington, VA, **156**
Columbia River Park, Columbia, PA, **92**

Columbus Riverfront Office Building, Columbus, GA, **118**
The Commons at Atlantic Station, Atlanta, GA, **124**
Cool Springs Mixed-Use Development, Franklin, TN, **152**
Court Street West Specific Plan, San Bernardino, CA, **218**
Crocker Park, Westlake, OH, **186**
Culinary Institute of America Anton Plaza, Hyde Park, NY, **94**

Dart CBD Transit Mall, Dallas, TX, **242**
Deep Blue Plaza, Hangzhou, China, **134**
Del Mar Station, Pasadena, CA, **182**
Discovery Communications Headquarters, Silver Spring, MD, **252**
Domus Office Tower, Hallandale Beach, FL, **232**
Downtown Detroit YMCA, Detroit, MI, **250**
Downtown Long Beach Visioning, Long Beach, CA, **220**
Downtown Projects, Seattle, WA, **23**
Downtown Silver Spring, Silver Spring, MD, **149**

Eastside, Richardson, TX, **51**

The Ferry Building Marketplace, San Francisco, CA, **42**
Ford's Landing, Alexandria, VA, **270**
Fudan Crowne Plaza Hotel, Shanghai, China, **200**

Galerias Hipodromo, Tijuana, Mexico, **219**
The Garden, Taichung, Taiwan, **222**
Gateway Center, Los Angeles, CA, **110**
Georgia Aquarium, Atlanta, GA, **288**

The Harbor, Rockwall, TX, **212**
Hard Rock Hotel, Chicago, IL, **166**
Highmark Data Center, Harrisburg, PA, **238**
Hollywood & Highland, Los Angeles, CA, **112**

Ibn Battuta Mall Phase II, Dubai, UAE, **78**

The Joffrey Tower, Chicago, IL **64**

Kincora, Loudoun County, VA, **84**

Lake Bluff Tower, Milwaukee, WI, **26**
Lane Field, San Diego, CA, **132**
Las Colinas Entertainment, Irving, TX, **76**
Lititz Watch Technicum, Lititz, PA, **90**
The Lofts at Stapleton, Denver, CO, **71**

Magnolia, Charleston, SC, **114**
MB Financial Bank, Chicago, IL, **60**
Melrose Triangle, West Hollywood, CA, **223**
Mercato, Bend, OR, **47**
Mercer Squar, Dallas, TX, **53**
Midtown Reston Town Center, Reston, VA, **154**
Museum Place, Fort Worth, TX, **127**

Nationwide Arena District, Columbus, OH, **190**
New Jersey Urban Parks Master Plan Competition, Trenton, NJ, **248**
New Jiang Wan Cultural Center, Shanghai, China, **234**
New River at Las Olas, Fort Lauderdale, FL, **109**
North Bank Park, Columbus, OH, **189**
North County Regional Education Center, San Marcos, CA, **131**

Odeon Union Square, San Francisco, CA, **174**
Other Urban Projects, Los Angeles, Sacramento, San Jose, CA; Reno, NV, **24**
Oxbow Public Market, Napa, CA, **46**

Palladium, Birmingham, MI, **142**
Palmolive Building, Chicago, IL, **62**
Park Kingsbury, Chicago, IL, **168**
Paxton Walk, Las Vegas, NV, **206**
The Pearl District, Portland, OR, **18**
Piemonte at Ontario Center, Ontario, CA, **36**
Plaza Norte, Santiago, Chile, **286**
Ponto Beachfront Village, Carlsbad, CA, **262**
Portico, San Diego, CA, **260**
Port Imperial, West New York, NJ, **158**
Providence 2020, Providence, RI, **244**

Redmond City Hall, Redmond, WA, **194**
Rio Nuevo, Tucson, AZ, **184**
Riverside Plaza, Riverside, CA, **34**
The Robert Redford Building for the Natural Resources Defense Council (NRDC), Santa Monica, CA, **178**
Rocketts Landing, Richmond, VA, **88**

Schenley Plaza, Pittsburgh, PA, **246**
Seattle Premium Outlet Center, Tulalip, WA, **38**
Shenzhen Bay Seafront Urban Design, Shenzhen, China, **150**
The Shops at Highland Village, Highland Village, TX, **214**

Sofitel JJ Oriental Hotel, Shanghai, China, **133**
Somerset Collection South Renovations, Troy, MI, **140**
Sonoma Mountain Village, Rehnert Park, CA, **170**
SouthPark Mall, Charlotte, NC, **146**
Sugarland Town Center Expansion, Sugarland, TX, **216**
Sustainable Neighborhoods, Fort Belvoit, VA; Philadelphia, PA; Washington, DC, **278**
Sustainable Urbanism, Los Angeles, CA, **280**
Sustainable Urbanism, Washington, DC, **274**

Technology Square, Georgia Institute of Technology, Atlanta, GA, **284**
Terminus, Atlanta, GA, **120**
TODs and Town Centers, Rockville, MD; Beaverton, OH; Loudoun County, VA, **276**
Town Center of Virginia Beach, Virginia Beach, VA, **82**
The Towne of Seahaven, Panama City Beach, FL, **148**
The Townlofts at Stapleton, Denver, CO, **71**
Towson Town Center, Towson, MD, **74**
Treasure Island Master Plan 2005, San Francisco, CA, **48**
The Triangle, Austin, TX, **56**
Triangle Town Center, Raleigh, NC, **287**
Tuck-Under Avenue Row Houses at Stapletonnn, Denver, CO, **72**

U.S. Epperson/Lynn Insurance Group Corporate Headquarters, Boca Raton, FL, **228**

The Village of Rochester Hills, Rochester Hills, MI, **138**
Visteon Village Corporate Headquarters, Van Buren Township, MI, **254**

Washington Convention Center, Washington, DC, **282**
Watermarke, Irvine, CA, **202**
West Highlands, Atlanta, GA, **126**
Workhouse Art Center at Lorttton, Lorton, VA, **266**
WRIT Rosslyn Center, Arlington, VA, **269**

X/O, 1712 South Prairie Avenue, Chicago, IL, **162**

Zhangilang Semiconductor Research Park, Phase II, Shanghai, China, **198**
Zhejiang Fortune Financial Center, Hangzhou, China, **196**